STABILITY OF ADAPTIVE SYSTEMS

The MIT Press Series in Signal Processing, Optimization, and Control
Alan S. Willsky, editor

1. *Location on Networks: Theory and Algorithms,* Gabriel Y. Handler and Pitu B. Mirchandani, 1979
2. *Digital Signal Processing and Control and Estimation Theory: Points of Tangency, Areas of Intersection, and Parallel Directions,* Alan S. Willsky, 1979
3. *Stability and Robustness of Multivariable Feedback Systems,* Michael George Safonov, 1980
4. *Issues in the Implementation of Digital Feedback Compensators,* Paul Moroney, 1983
5. *Theory and Practice of Recursive Identification,* Lennart Ljung and Torsten Söderström, 1983
6. *Approximation and Weak Convergence Methods for Random Processes, with Applications to Stochastic Systems Theory,* Harold J. Kushner, 1984
7. *Control System Synthesis: A Factorization Approach,* M. Vidyasagar, 1985
8. *Stability of Adaptive Systems: Passivity and Averaging Analysis,* B. D. O. Anderson, R. R. Bitmead, C. R. Johnson, Jr., P. V. Kokotovic, R. L. Kosut, I. M. Y. Mareels, L. Praly, and B. D. Riedle, 1986

STABILITY OF ADAPTIVE SYSTEMS: PASSIVITY AND AVERAGING ANALYSIS

B. D. O. Anderson, R. R. Bitmead, C. R. Johnson, Jr.,
P. V. Kokotovic, R. L. Kosut, I. M. Y. Mareels,
L. Praly, and B. D. Riedle

(in alphabetical order)

The MIT Press
Cambridge, Massachusetts
London, England

PUBLISHER'S NOTE

This format is intended to reduce the cost of publishing certain works in book form and to shorten the gap between editorial preparation and final publication. Detailed editing and composition have been avoided by photographing the text of this book directly from the author's prepared copy.

This book was printed and bound in the United States of America.

Library of Congress Cataloging-in-Publication Data

Stability of adaptive systems.

(The MIT Press series in signal processing, optimization, and control; 8)

Includes bibliographies and index

1. Adaptive control systems. 2. Stability. I. Anderson, Briän, D. O., II. Series.

TJ217.S735 1986 629.8'36 86-2705

ISBN 0-262-01090-9

AUTHOR AFFILIATIONS

Brian D.O. Anderson	*Department of Systems Engineering* *Australian National University* *Canberra, Australian Capital Territory, AUSTRALIA*
Robert R. Bitmead	*Department of Systems Engineering* *Australian National University* *Canberra, Australian Capital Territory, AUSTRALIA*
C. Richard Johnson, Jr.	*School of Electrical Engineering* *Cornell University* *Ithaca, New York, USA*
Petar V. Kokotovic	*Department of Electrical and Computer Engineering* *Coordinated Science Laboratory* *University of Illinois, Urbana, Ill., USA*
Robert L. Kosut	*Integrated Systems, Inc., Palo Alto, CA* *Department of Electrical Engineering* *Stanford University, Stanford, CA, USA*
Iven M.Y. Mareels	*Department of Systems Engineering* *Australian National University* *Canberra, Australian Capital Territory, AUSTRALIA*
Laurent Praly	*Centre d'Automatique et Informatique* *Ecole des Mines* *Fontainebleau, FRANCE*
Bradley D. Riedle	*Department of Electrical and Computer Engineering* *Coordinated Science Laboratory* *University of Illinois, Urbana, Ill., USA*

Support Acknowledgements

Research School of Physical Sciences, Australian National University, (BDOA, RRB, CRJ, RLK, PVK, IMYM); *Radio Research Board, Australia* (RRB)
National Science Foundation: Grants ECS-85-06149 (CRJ), ECS-83-11851 (PVK, BDR, LP); *and US/Australia Coop. Sci. Program, Grant* INT-85-13400 (CRJ, RLK); *Department of Science, Australia,* (RRB).
Air Force Office of Scientific Research, Dir. of Aerospace Sci., Contract F49620-85-C-0094, *and Dir. of Math. and Info. Sci., Contract* F49620-84-C-0085 (RLK)
Joint Services Electr. Program, Contract N00014-84-C-0149 (PVK, BDR, LP)
National Fund for Scientific Research, Belgium (IMYM)
Groupe de Recherche Coordonnee-Systemes Adaptatifs en Robotique, Traitement du Signal et Automatique - du CNRS (LP)

CONTENTS

SERIES FOREWORD

The fields of signal processing, optimization, and control stand as well-developed disciplines with solid theoretical and methodological foundations. While the development of each of these fields is of great importance, many future problems will require the combined efforts of researchers in all of the disciplines. Among these challenges are the analysis, design, and optimization of large and complex systems, the effective utilization of the capabilities provided by recent developments in digital technology for the design of high-performance control and signal-processing systems, and the application of systems concepts to a variety of applications, such as transportation systems, seismic signal processing, and data communication networks.

This series serves several purposes. It not only includes books at the leading edge of research in each field but also emphasizes theoretical research, analytical techniques, and applications that merit the attention of workers in all disciplines. In this way the series should help acquaint researchers in each field with other perspectives and techniques and provide cornerstones for the development of new research areas within each discipline and across the boundaries.

The analysis of adaptive systems is a particularly appropriate topic for a book in this series, given its relevance to so many problems of interest to researchers in signal processing and systems and control and given the considerable intensification of interest in and research on this subject in recent years. As the reader is probably aware, the field of adaptive systems is filled with a seemingly endless variety of methods, with assumptions and conditions whose fundamental significance may be far less apparent than the less than completely satisfying fact that they seem to make things work, and with controversy concerning the performance of these methods in practical applications. Given this background, the nonspecialist might, with some justification, view a book with eight authors as a fitting contribution to the cacophony someone once referred to as the Grand Bazaar of adaptive systems. This nonspecialist would be wrong. Brian Anderson, Robert Bitmead, C. Richard Johnson, Petar Kokotovic, Robert Kosut, Iven Mareels, Laurent Praly, and Bradley Riedle's book *Stability of Adaptive Systems: Passivity and Averaging Analysis* is a most impressive achievement and an extremely important and welcome addition to the literature on adaptive systems.

As the authors point out, this book is somewhere between a research monograph and a textbook on stability of adaptive systems. The book actually deserves both descriptors. Without doubt the results and methods presented in the book represent extremely important contributions to the theory of adaptive systems. As is often the case, the most difficult theoretical problems involving signals and systems center on issues related to making a methodology work in practice. In the case of adaptive systems this translates into the difficult problem of ensuring stable operation of an adaptive system in a real environment. The authors have taken this problem square on and have produced a most impressive theory of robust stability of adaptive systems answering many of the questions and challenges raised in recent years. This achievement certainly makes this book an important one for researchers and practitioners in adaptive systems.

What is perhaps equally impressive about this book is the cohesiveness of the treatment, a fact that is deserving of praise in a single-author textbook on a well-documented topic and a source of amazement when there are as many authors as there are words in the title and the topic is as recent in its development as this one is. This book can most definitely be used as the basis for a course on adaptive systems, and furthermore, the fundamental nature of the results presented should ensure this book a long life.

Starting from a basic total stability theorem and several of the fundamental principles underlying much of stability analysis -- passivity, small gain theorems, and averaging methods -- the authors develop a comparatively (and laudably!) small number of extremely general and powerful stability results for adaptive systems. These few results provide considerable insight into the mechanisms of stability and instability in adaptive systems and into the central role played by conditions on system positivity and richness of reference signals. These few results then provide the basis for analyzing the stability and robustness of a wide variety of adaptive systems. Through these analyses, the notes and references provided at the ends of chapters and at numerous other points in the development, and the most welcome inclusion of a design example and the remarkable commentary that accompanies and follows it, the reader develops an appreciation not only for the theory developed in this book but also for the nature of adaptive systems and the rich "folklore" and tricks of the trade for which the methods and insights developed in this book provide a theoretical interpretation and a rational basis. Without question the sophisticated development in this book presents a challenge to the reader, and mastery of the material requires a commitment of some time and energy. However, as the reader progresses -- or, should I say, bounds -- from chapter to chapter of *Stability of Adaptive Systems*, I think that he or she will find more than an ample return on the investment.

Alan S. Willsky

PREFACE

The tempting vision of engineering designs and algorithms which adapt to an ever changing environment is being pursued by many researchers trying to achieve convergence not only of adjustable parameters, but also of their diverse theories. This volume is itself a convergence result: its eight authors have converged to a set of passivity and averaging concepts organized into a pedagogical sequence of chapters, rather than a collection of papers. Each chapter is the result of a cooperative draft-review-rewrite effort, based almost exclusively on the current research of the authors. Notes and References stress this fact and do not attempt to overview all other important works.

This book offers insights into the behavior of adaptive systems from the viewpoint of stability theory, employing both input/output and Lyapunov stability concepts. While passivity theory shows input/output stability, the method of averaging significantly relaxes the restrictive passivity conditions. The stability analysis is extended to reveal the causes and mechanisms of instability and to suggest means to counteract them. The emphasis is on methodology and basic concepts, rather than on details of adaptive algorithms. Hence, the adaptive algorithms considered are of the simplest form, clearly exhibiting common properties of more elaborate schemes. Simultaneous treatment of continuous and discrete-time systems stresses the similarity of the results.

The presentation starts with the concepts of the tuned system, linearization, and total stability in Chapter 1, time-varying linear analysis in Chapter 2, linear averaging theory in Chapter 3, and then extends them to nonlinear continuous systems in Chapter 4 and nonlinear discrete systems in Chapter 5. Chapter 6 offers an extensive illustrative example with comments and a suggestion for algorithm modification based on regressor filtering. Simple examples are presented throughout to illustrate the uses and limitations of the theory.

The perceived place of this book in the adaptive control literature is somewhere between a purely research monograph and a basic pedagogical exposition. This is because we develop a ***robustness theory*** for adaptive systems without all the specifics of the algorithms, which are the province of the textbooks by Landau (1979), Ljung and Soderstrom (1983), and Goodwin and Sin (1984). These texts show that, under some ***ideal modeling assumptions,*** most of the basic algorithms possess global convergence properties. On the other hand, the more specialized monographs of Egardt (1979) and Ioannou and Kokotovic (1983) demonstrate that such global properties are not robust and can be lost due to disturbances, lack of excitation, violation of positive realness (passivity) conditions, unmodeled dynamics and similar modeling imperfections always present in actual systems. These robustness issues are also addressed in this book, but with new, sharper tools which, hopefully, have produced clearer results. With these tools it was possible first to decipher many of the cryptic notions of adaptive control -- strictly positive real conditions, persistence of excitation, etc. -- and then to replace them with more robust signal dependent positivity conditions which remain valid, at least locally, in the presence of disturbances and unmodeled dynamics. Thus, while the intent and content of our work here is more in line with the monographs of Egardt (1979) and Ioannou and Kokotovic (1983), we pace our development to suit a less specialized audience.

Perhaps the most remarkable aspect of this book is the number of its authors -- a direct result of the friendship and technical collaboration established over an extended period. While, unfortunately, some of the authors have never met face to face with all of the others, the emergence of this manuscript was assisted by travel exceeding a quarter of a million miles.

In this epoch of formation of national research centers, it might sound a discordant note that cooperative research can flourish at disperse locations. The genesis of the whole concept was in August 1984 in Bozeman, Montana, USA -- *home to none of us* -- but made so by the warm hospitality of Don Pierre of Montana State University. Subsequent venues and approximate dates, where intensive joint work on the book was part of an otherwise hectic schedule, were as follows: Urbana, Illinois (October '84 and December '85); Canberra, Australian Capital Territory (November '84, May '85 and February '86); Ithaca, New York (August-December '84, June '85); Las Vegas, Nevada (December '84); Los Angeles, California (May '85); Boston, Massachusetts (June '85); Bozeman, Montana (August '85); Lund, Sweden (August '85); Palo Alto, California (November '85); Fort Lauderdale, Florida (December '85).

In this process of wandering we inconvenienced not only our families, but also many friends and colleagues imprudent enough to share our enthusiasm for this project. This way of thanking them for support is inadequate but the volume of writing down all the gratitudes would be overwhelming. Caught in the web were our editor, Alan Willsky, and publisher, Frank Satlow, who, while continuously puzzled by the phenomenon of eight authors writing a book, managed to produce it without delay. This accomplishment was made possible by the expert typing of Patricia Krokel of Stanford University who stoically endured the frenzy of eightfold correcting and proofreading. Her patient assistance and perserverance have been irreplaceable.

And, to conclude with an even more remarkable facet of this project: at its completion all friendships and the enthusiasms for the research in adaptive systems not only survived, but have continued to grow.

"It is an error to imagine that evolution signifies a constant tendency to increased perfection. That process undoubtedly involves a constant remodeling of the organism in adaptation to new conditions; but it depends on the nature of those conditions whether the direction of the modifications effected shall be upward or downward."

-- T.H. Huxley

STABILITY OF ADAPTIVE SYSTEMS

Chapter 1

ROBUST STABILITY FORMULATION

1.1 INTRODUCTION

Adaptive systems are a response to engineering design problems arising from uncertainty. In the representative areas of adaptive control, communications, and signal processing, uncertainty means an imprecise knowledge of the current system. The aim of adaptation is to provide on-line modification of system behavior in response to current measured performance. Thus, the formulation of adaptive systems involves concepts of performance measures, adjustment rules, and desired objectives, together with an appreciation of the role of uncertainty in describing the system.

In characterizing the effectiveness of adaptive methods one is led immediately to questions concerning the nature of the response of the adaptive system vis-a-vis a hypothetical nonadaptive system designed according to the same objectives but with complete system knowledge, or more properly, with a much reduced level of uncertainty. The primary question raised by this characterization is whether the adaptive system achieves a performance close to that of the hypothetical "ideal." Thus, *stability* arises as an issue in adaptive systems and deals with the adaptive system per se or its deviation from the platonic ideal. It describes the signals in the system and their asymptotic properties such as boundedness or convergence. When we discuss *robust stability* in this framework, we include into our assessment the preservation of these stability notions when a type of uncertainty is allowed to be present in the system which is not explicitly accounted for in the adaptive mechanism. Specifically, we will be concerned with a system or plant which is variously characterized by *structured* parametric uncertainty and *unstructured* modeling error. The adaptive system responds directly to the structured uncertainties by the adjustment of selected system parameters. By its nature, unstructured uncertainty is not modelled in the analysis, and so must be quantified and introduced in a different fashion.

Historically, the developments in adaptive systems -- essentially, parameter adaptive systems -- have concentrated on the issues of maintaining good bounds between adaptive performance and ideal nonadaptive performance ("reference model") in the absence of unstructured uncertainty. For the more complicated areas such as adaptive control, it has not been possible to account fully for the behavior with unstructured uncertainty because of the occurrence of nonlinear dynamical equations whose solutions have been difficult to qualify. The emphasis in this area has therefore been primarily to develop convergence theories to demonstrate the ability of adaptive schemes to overcome structured parametric uncertainty. This has led to a class of important results presenting global convergence and asymptotic optimality. With the inclusion of additional unstructured effects it proves no longer possible to preserve these global results in a broad range of situations. Consequently, the development of robust stability theory for adaptive systems in its most general form must take place in a local context, although some global results are still achievable in some instances.

Our central objective in this treatise is to develop, in as cogent a manner as possible, a theory of the robust stability of adaptive systems. Our thrust will primarily be directed towards a qualitative understanding of the mechanisms of robust local stability so that questions concerning the suitability of adaptation in certain practical circumstances can be addressed. More quantitative characterizations then are derived subsequently from this basis. In an attempt to promulgate our creed most widely, we stress the pedagogical aspects of the theory at the same time as deriving new results and concentrate on developing the conceptual and intuitive fundamentals. This focus comes at the expense of an exhaustive analysis of many specific variants of adaptive algorithms and of the presentation of the most flexible but less transparent supporting results. Typically, sophisticated extensions possess an identical conceptual basis.

Summary of Chapters: In the rest of this chapter we provide a derivation of the general equations describing adaptive systems and an introduction of our main tool for the underlying theory of robust stability in the face of unstructured uncertainty -- the Total Stability Theorem. The specific approach taken is to describe the evolution of the signals in the adaptive system in terms of their deviation from nominal "tuned" values which exemplify the ideal design discussed above. Thus, we do not consider the nonlinear dynamical equations of the adaptive system directly, but rather the equations of an error system. These equations are still nonlinear but, subject to certain regularity conditions, do admit a linearization about zero error. From the Total Stability Theorem, exponential stability of the linearized system implies local exponential stability of an associated nonlinear system. This, in turn, establishes conditions for the local boundedness of the state of the complete nonlinear error system in terms of restrictions on the magnitude of initial conditions and unstructured error signals. This presentation is in general terms of Banach spaces, fixed point theorems, operator notations, etc., and supports the body of our work where we tie our analysis more closely to standard adaptive methods, at which stage we utilize state-variable descriptions more than operators. We conclude Chapter 1 with the development of the standard form of the dynamical equations governing the *adaptive error system,* including the underlying linearized form, i.e., the *linear error system*.

We next turn to the problem of deriving sufficient conditions for the exponential stability of the linear error system. This is done in Chapters 2 and 3 using the separate techniques of passivity analysis and averaging, respectively. In these analyses, the effect of the unstructured uncertainty is reflected in imprecision of the knowledge of the linear operator appearing in the linearized equations. In Chapter 2 the stability of the linear equation is analyzed using passivity and small gain methods -- these encompass the renowned strictly positive real (SPR) condition of adaptive systems, and reconstitute this positivity requirement to more general signal dependent operator forms. This *signal dependent positivity condition,* also referred to as a *dominant richness condition,* enlarges the notion of persistence of excitation of input signals to the adaptive system, and is a major theme of the book. For example, when the unstructured uncertainty reflects high frequency dynamics, the condition is composed of a low-frequency positivity requirement on the operator coupled with a nondegeneracy and dominant low-pass stipulation on the signals within the adaptive system.

In Chapter 3, the alternative techniques of averaging and explicit time-scale separation further develop specifications of sufficient conditions for exponential stability of the linear equation. This approach makes clear the role played in the theory by both the dominant richness assumption and slow adaptation rate. These techniques produce stability and instability conditions for the linear equation and provide the most transparent statement of the dominant richness condition in terms of an average positivity condition of signal products, i.e., a *signal dependent "average SPR" condition.* The theory of averaging is extended in this chapter to include signals with *sample averages.* Illustrative examples include the linearized analysis of an output error algorithm.

Chapters 4 and 5 contain derivations of the specific error equations which arise in particular, but typical, examples of adaptive identification and control: equation error, output error, and model reference adaptive control. The continuous-time case is studied in Chapter 4, while Chapter 5 contains the discrete-time case. Both chapters provide a local stability analysis of the adaptive error systems. The analyses are based on the Total Stability Theorem of Chapter 1 and the method of averaging and time-scale decomposition of Chapter 3. These analyses are a demonstration of the power and utility of the theory and are principally designed to specify sufficient conditions for the robust stability of these schemes in realistic engineering terms, using signal properties in addition to system properties. The aim of specifying such conditions is to clarify the requirements for answering questions about the feasibility or suitability of using adaptive control in certain circumstances. The clear message is that a response to such an inquiry requires restrictions on the system as well as on its signal environment.

After the first five chapters, our theory development is complete. In Chapter 6 an extensive control example is presented to display both the utility and the limitations of the analysis. Perhaps more important, the synergism of analytical theory and practical requirements is displayed by the ability of the theoretical developments to suggest practical modifications to adaptive control algorithms which significantly enhance their performance. The modification considered here consists of including appropriate filters in certain parts of the adaptation procedure in such a way as to enhance the satisfaction of the sufficient conditions for robust stability. The book concludes with a description of practical implications of these analytical techniques for adaptive systems in applications. In particular, we return to our stated aims above to assess the level to which we have satisfied our goals and to cautiously presage their impact on the theory as a whole and on potential applications.

1.2 ADAPTIVE SYSTEMS

Adaptation in systems theory describes a design methodology based on the on-line estimation of desired quantities to compensate for unknown fixed or slowly time-varying plant characteristics. In adaptive identification the aim is to derive a model of the plant used for certain purposes. For example; adaptive equalization is directed to approximating the inverse of a plant, typically a communication channel; adaptive control is more ambitious and strives to generate a feedback controller for the unknown plant and, further, possibly to do so in accordance with clear control objectives of standard type.

There is a wealth of subtle taxonomy linking notions of adaptation, learning, self-optimization, artificial intelligence, pattern recognition, etc. ..., but, to define our territory clearly, we shall be considering *parameter adaptive systems,* that is, the characterization of the unknown but desired quantity of the design is a finite-dimensional vector of real parameters, denoted by θ, which is to be estimated on-line from plant signals. The parameters need not be chosen explicitly to produce a plant model but, as is evidenced above, may be selected for some related objective using the plant signals. We therefore consider the interconnection of two causal dynamical systems: the plant which generates the signals, possibly in a fashion depending upon the current parameter estimate value, and the adaptation mechanism which adjusts the parameter estimates themselves. For the moment we shall leave in abeyance the question of the precise nature of the plant system and address that of the adaptation procedure's formulation.

Coupled with finite-dimensional parameter estimation is the need to represent state information of the modelled system based on measured plant signals. This finite-dimensional model state will be called *the regressor vector signal,* by analogy with the formulation of linear regressor modelling problems, and will be denoted by $\phi(t)$. The role of the adaptive estimator is to take the regression vector signal $\phi(t)$, past parameter estimates $\hat{\theta}(t)$, and an error signal $e(t)$ and to update or compute a new value of the parameter estimate. While $\phi(t)$ represents a model state derived from plant information, the error signal $e(t)$ is an instantaneous performance measure for the parameter estimator. Typically $e(t)$ is computed as the difference between a signal generated using $\hat{\theta}(t)$ and a reference objective function provided externally or derived from plant measurements. The particular form of the adaptation mechanism will be specified later, but most often it is composed of an approximate gradient descent or Newton-Raphson iterative procedure familiar from statistical point estimation for the minimization of a mean-squared error criterion with linear regression models. In this latter context, another coefficient becomes important -- ϵ, the adaptation algorithm gain or stepsize -- which dictates the speed of adaptation.

We now have identified signals in the adaptive system. If we define the collection of exogenous signals impinging on our system as $w(t)$, which includes external inputs, reference signals, disturbances, and noise, then the complete adaptive system may be described as the interconnection of:

The Regressor Subsystem which takes as inputs $w(t)$ and $\hat{\theta}(t)$ to produce the signal $\phi(t)$. Because $\hat{\theta}(t)$ is meant to represent a parameter upon which the instantaneous closed-loop plant may depend we describe this subsystem as an operator on w

$$\phi = H_{\phi w}(\hat{\theta})w \tag{2.1}$$

with a parametric dependence on $\hat{\theta}$. The regressor subsystem usually involves the plant and may include a feedback controller, as in adaptive control, or a feedback model of the plant, as in output error identification, as well.

The Error Subsystem which takes as inputs $w(t)$, $\phi(t)$, and $\hat{\theta}(t)$ to produce the error signal $e(t)$ for the adjustment in the adaptation. Again we note the parametric dependence of e on $\hat{\theta}$ in our notation and also use (2.1) to eliminate the explicit dependence on ϕ to write

$$e = H_{ew}(\hat{\theta})w \tag{2.2}$$

The Adaptation Subsystem which uses e and ϕ to generate parameter estimates $\hat{\theta}$. Clearly, there is no parametric dependence on $\hat{\theta}$ here since $\hat{\theta}$ is part of the state of this subsystem. Because of distinctions to be drawn later we write this algorithm for updating $\hat{\theta}$ as

$$\hat{\theta} = \Omega(e,\phi) \tag{2.3}$$

We could also include a parametric dependence in (2.3) of Ω on ϵ, the stepsize coefficient, and on $\hat{\theta}_0$, the initial parameter estimate.

This decomposition of the adaptive system into these three blocks is nonstandard in that the plant is no longer explicitly visible and the error computation is generalized from a simple prediction error. This structure is chosen because it highlights our emphasis on signal properties in adaptation and the desired general characteristics of the operators connecting these signals rather than relying too heavily on plant/controller structures and parameterizations.

We may write the composite adaptive system as

$$\begin{bmatrix} e \\ \phi \end{bmatrix} = \begin{bmatrix} H_{ew}(\hat{\theta}) \\ H_{\phi w}(\hat{\theta}) \end{bmatrix} w = H(\hat{\theta})w \tag{2.4a}$$

$$\hat{\theta} = \Omega(e,\phi) \tag{2.4b}$$

This is depicted in Figure 1.1.

The topic of study in this book will be the dynamic behavior of the adaptive system (2.4) and the relationship between this behavior and the specification and execution of the adaptive problems in engineering. Our aim will be to derive conditions for the boundedness of e and $\hat{\theta}$ in return for constraints on w, the initial condition values, and the operators above. In general (and particularly in adaptive control) (2.4) is a nonlinear set of coupled equations in spite of the fact that the operators are, nominally linear, i.e., $H(\hat{\theta})$ is linear for fixed $\hat{\theta}$ -- the nonlinearity effectively is due to the parametric influence of $\hat{\theta}$. In certain circumstances it is

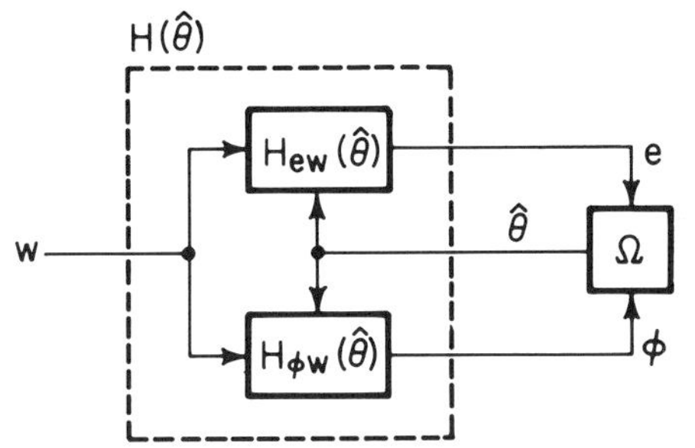

Fig. 1.1 The adaptive system.

possible to prove global convergence of $\hat{\theta}$ and boundedness of ϕ and thereby to establish global stability properties for (2.4). This will be amplified later in this chapter. However, the assumptions on which this global stability is based are not robust and so we concentrate on developing local stability results for (2.4) which are robust in the sense that small perturbations of plants, algorithms, operators, external signals, initial conditions provide small perturbations of internal signals.

The route we take to study the local stability properties of (2.4) is to use the techniques of linearization and total stability. Linearization dictates that, subject to smoothness and regularity conditions of (2.4), we may construct a linear set of equations from (2.4) whose stability properties are identical to the local stability properties of (2.4). Total stability theory allows us to derive conditions for the robustness of local stability results to guarantee maintenance of local exponential stability or bounded-input/bounded-output stability in the presence of small perturbations.

1.3 ERROR SYSTEMS, LINEARIZATION, AND TOTAL STABILITY

With the adaptive system structure defined by (2.4) we now introduce the theoretical tools required to develop a robust stability theory for the adaptive system. In the next section we abide by our operator notation to develop the conceptual framework of the adaptive error system and the linearization about nominal trajectories. To complete the development of the stability conditions we then move on to present the total stability theorem which links the local stability properties of the nonlinear system to those in perturbed linearized form. The total stability theorem will be stated both in the operator form and in the more familiar differential equation form foreshadowing our subsequent switch in notation when more detailed structure is imposed on the adaptive problems considered. In particular, state-variable equations will be introduced along with intuitively pleasing mixed forms.

1.3.1 Conceptual Framework for Linearization

Our analysis in this section will proceed by recasting the adaptive system equations (2.4) in an incremental form so that their explicit subjects are the deviations of e, ϕ, $\hat{\theta}$ from nominal, *tuned* values e_*, ϕ_*, θ_*. The technique then will be to write down linearized versions of the incremental equations whose solution properties reflect the local properties of solutions of (2.4). In this way, qualitative statements about (2.4) and its solutions may be made. The mechanics of this analysis will be via total stability and linearization results and these will be introduced later as needed. We consider first the question of the error system.

Let $e_*(t)$, $\phi_*(t)$, $\theta_*(t)$ be specified functions of time. Then we may rewrite the adaptive system (2.4) in terms of error signals

$$\tilde{e}(t) = e_*(t) - e(t)$$

$$\tilde{\phi}(t) = \phi_*(t) - \phi(t) \tag{3.1}$$
$$\theta(t) = \theta_*(t) - \hat{\theta}(t)$$

to yield the incremental description

$$\tilde{e} = H_{ew}(\theta_*)w - H_{ew}(\hat{\theta})w + \delta_e$$
$$\tilde{\phi} = H_{\phi w}(\theta_*)w - H_{\phi w}(\hat{\theta})w + \delta_\phi \tag{3.2}$$
$$\theta = \Omega(e_*, \phi_*) - \Omega(e, \phi) + \delta_\theta$$

where

$$\delta_e = e_* - H_{ew}(\theta_*)w$$
$$\delta_\phi = \phi_* - H_{\phi w}(\theta_*)w \tag{3.3}$$
$$\delta_\theta = \theta_* - \Omega(e_*, \phi_*)$$

Clearly, the incremental description (3.2) is identical to (2.4).

Equations (3.2) are nonlinear in general but, if the operators $H_{ew}(\theta)$, $H_{\phi w}(\theta)$, $\Omega(e,\phi)$ are locally differentiable (as operators) with respect to their arguments in a ball B centered on e_*, ϕ_*, θ_*, i.e.,

$$B = \{e, \phi, \theta : \|e_* - e\| \le r_e,\ \ \|\phi_* - \phi\| \le r_\phi,\ \ \|\theta_* - \theta\| < r_\theta\} \tag{3.4}$$

then we may act as if there were linear operators

$$\frac{\partial H_{ew}}{\partial \theta}\Big|_{\theta_*} w,\quad \frac{\partial H_{\phi w}}{\partial \theta}\Big|_{\theta_*} w,\quad \frac{\partial \Omega}{\partial e}\Big|_{e_*,\theta_*},\ \text{and}\ \frac{\partial \Omega}{\partial \phi}\Big|_{e_*,\theta_*}$$

so that, in B, (3.2) becomes identical to

$$\tilde{e} = \left(\frac{\partial H_{ew}}{\partial \theta}\Big|_{\theta_*} w\right)\theta + \delta_1 + \delta_e$$
$$\tilde{\phi} = \left(\frac{\partial H_{\phi w}}{\partial \theta}\Big|_{\theta_*} w\right)\theta + \delta_2 + \delta_\phi \tag{3.5}$$
$$\theta = \frac{\partial \Omega}{\partial e}\Big|_{e_*,\theta_*} e + \frac{\partial \Omega}{\partial \phi}\Big|_{e_*,\theta_*} \phi + \delta_3 + \delta_\theta$$

where δ_1, δ_2, δ_3 are bounded in norm by $K_1\|\theta\|^2$, $K_2\|\theta\|^2$, and

$K_3(\|\tilde{e}\|^2 + \|\tilde{\phi}\|^2)$ uniformly in B. We do not carry through the detailed derivation of (3.5), which involves the introduction of Frechet derivatives and some cumbersome machinery, since it will be subsumed shortly by more elegant but less transparent fixed point theorems. Nevertheless, the visible structure of (3.5) conveys much which underlies our whole approach.

Equations (2.4) describe the interconnection of the component subsystems of an adaptive scheme: the regressor system including the plant, the error system, and the parameter adaptation. Given nominal trajectories e_*, ϕ_*, θ_* for the interconnection signals of the system, it was then possible to rewrite (2.4) in its incremental form (3.2). It should be stressed that (3.2) is identical to (2.4) and that, so far, no assumption of smallness of deviations of signals had been introduced. The major intellectual leap comes in recognizing that, if the operators are sufficiently smooth functions of their arguments in a ball B, (3.2) and hence (2.4) are identical to (3.5), which consists of a collection of linear operator equations with additive driving terms of two kinds -- a priori bounded functions given by (3.3) and terms which are uniformly quadratically bounded in B. Thus (3.5) describes, at least formally, a linearization of (3.2) or (2.4) about the signal values e_*, ϕ_*, θ_*.

We are immediately inspired to phrase several questions:

(1) How can (3.5) aid us in studying the properties of solutions of (2.4)?

(2) How should we choose the nominal signals e_*, ϕ_*, θ_* and what do they mean?

(3) How can these abstract operator-theoretic notions lead us to an understanding of the behavior of practical adaptive systems?

The first of these questions will be addressed shortly in the Total Stability Theorem. Effectively, if the linear operator in (3.5) is contractive, then, provided the δ signals are adequately bounded, there exists a unique, bounded solution of (3.5) and hence (2.4) in B. In state-space terms this corresponds to demanding exponential stability of the linear homogeneous part of (3.5) and then invoking bounded-input/bounded-state stability ideas. Thus existence and boundedness of solutions of (2.4) follows. The answer to the second question draws on that of the first: the nominal signals are chosen to enhance the requirements of the Total Stability Theorem, although it should be said that in some circumstances it is never expected that these signals actually are specified since our thrust in the book is primarily to develop qualitative sufficient theories. The final question is the subject of the book: to develop and translate the mathematical theory of robust adaptive systems to enhance its applicability in engineering problems. To this end, some explicit error models for adaptive systems are derived to the later part of this chapter which demonstrate a simple form.

1.3.2 Total Stability

We move on now to present the Total Stability Theorem in two manifestations -- operator form and state-space form. For brevity, we introduce the notation:

$$x = (\tilde{e}^T \ \ \tilde{\phi}^T \ \ \theta^T)^T \tag{3.6}$$

$$x_* = (e_*^T \ \ \phi_*^T \ \ e_*^T)^T \tag{3.7}$$

$$Q(x) = ((H_{ew}(\hat{\theta})w)^T \ (H_{\phi w}(\hat{\theta})w)^T \ \Omega(e,\phi)^T)^T \tag{3.8}$$

$$F(x) = Q(x_*) - Q(x) \tag{3.9}$$

$$\delta = (\delta_e^T \ \ \delta_\phi^T \ \ \delta_\theta^T)^T \tag{3.10}$$

In this notation (3.5) becomes

$$x = F(x) + \delta \tag{3.11}$$

The Total Stability Theorem relies on the contraction mapping principle of Banach and Cacciopoli. For completeness we state it here, but first we need some definitions.

If R is a subset of a Banach space of functions B, and T is an operator mapping $R \to B$, then T is a *contraction on* R if there is a constant $\sigma \in [0,1)$ such that

$$\|Tx - Ty\| \leq \sigma \|x - y\|, \quad \forall\ x,y \in R$$

The constant σ is called the *contraction constant for* T *on* R. A *fixed point* of $T: B \to B$ is a point $x \in B$ such that $x = Tx$.

Theorem 1.1: Contraction Mapping Principle

If R is a closed subset of a Banach space B and $T: R \to R$ is a contraction on R, then T has a unique fixed point in R.

□

This statement of the Contraction Mapping Principle can be found in Hale (1980). We now have:

Theorem 1.2: Total Stability (Operator Form)

Let B_r denote an r-bounded ball in the Banach space B, i.e.,

$$B_r = \{x \in B: \|x\| \leq r\}$$

Consider the system

$$x = F(x) + \delta \tag{3.12}$$

where

(i) $F(0) = 0$

(ii) $\exists\ r > 0$ and $\sigma \in [0,1)$ such that

$$\|F(x) - F(y)\| \leq \sigma \|x - y\|, \quad \forall\ x,y \in B_r$$

(iii) $\delta \in B_{(1-\sigma)r}$

Then, (3.12) has a unique fixed point in B_r.

□

Discussion

This theorem is a direct application of the Contraction Mapping Principle, Theorem 1.1. The term *Total Stability* stems from the fact that it can be viewed as providing sufficient conditions to preserve a fixed point: a small perturbation of a contractive operator does not destroy the existence of a fixed point. By *point* here we mean an element of a normed function space and total stability implies small perturbation of the functions from small perturbations of the operators. The theorem can be applied directly to (3.5), since (3.12) has the same form. Condition (ii) calls for F in (3.12) to be a contraction on B_r with contraction constant σ. In terms of (3.5) we can write

$$F(\tilde{x}) = F_{\tilde{x}}(0)\tilde{x} + \Delta$$

where $F_{\tilde{x}}(0) = \partial F(0)/\partial\tilde{x}$ contains the "partials" listed after (3.4), and $\Delta = (\delta_1, \delta_2, \delta_3)^T$ is the vector of second order terms in the expansions of $F(\tilde{x})$ about $\tilde{x} = 0$, i.e., about $x = x_*$ as defined in (3.6)-(3.7). Observe that $F_{\tilde{x}}(0)$ is in general a linear time-varying integral operator, and hence, the contractiveness of F will be dependent on the stability of the linear map $u \to y$ defined by

$$y = F_{\tilde{x}}(0)y + u$$

which is equivalently expressed as

$$y = [I - F_{\tilde{x}}(0)]^{-1}u$$

We will see shortly, that the contractiveness of F is insured (for some small r) if the map $u \to y$ is exponentially stable. Hence, we can view the linear inhomogeneous incremental system

$$\tilde{x} = F_{\tilde{x}}(0)\tilde{x} + \delta$$

as the *linearization of* (2.4) *about the trajectory* x_*.

We also remark that in general the operator Ω (typically an integrator) has an infinite gain -- i.e., is not contractive -- however, the "closed loop" (3.2) or (3.5) can still be contractive due to the interaction of the different operators -- much in the same way as a negative feedback stabilizes an integrator.

The operator formalism allows us simultaneously to handle under one framework all manner of systems, e.g., continuous or discrete, linear or nonlinear, lumped or distributed, etc. However, as these operator results are also somewhat opaque, and because we will make extensive use of state-space representation of the adaptive systems at hand, we also give the corresponding Total Stability Theorem for state-space systems. We start with continuous-time systems described by ordinary differential equations, and where appropriate, the analysis will be repeated for discrete-time systems described by ordinary difference equations.

Theorem 1.3: Total Stability (State-Space Form)

Consider the ordinary differential equation

$$\begin{aligned} \dot{x} &= A(t)x+f(t,x)+g(t,x) \\ x(t_0) &= x_0 \in \mathbb{R}^n \end{aligned} \tag{3.13}$$

where $A(t)$, $f(t,x)$ and $g(t,x)$, for each fixed x in the ball $|x| \leq r$, are locally integrable functions of t, and $\forall\, |x_1| \leq r$, $\forall\, |x_2| \leq r$, and $\forall\ t \geq t_0$:

$$\begin{aligned} &\text{(A1)}\quad f(t,0) = 0 \\ &\text{(A2)}\quad |f(t,x_1)-f(t,x_2)| \leq \beta_1|x_1-x_2| \\ &\text{(A3)}\quad |g(t,x_1)| \leq \beta_2 r \\ &\text{(A4)}\quad |g(t,x_1)-g(t,x_2)| \leq \beta_2|x_1-x_2| \end{aligned} \tag{3.14}$$

If the unperturbed system

$$\dot{x} = A(t)x \tag{3.15}$$

is exponentially stable, i.e., if for some constants $a>0$ and $K\geq 1$ the state transition matrix $F(t_1,t_2)$ of (3.15) satisfies

$$|F(t_2,t_1)| \leq Ke^{-a(t_2-t_1)} \quad \forall\ t_2\geq t_1\geq t_0 \tag{3.16}$$

and if

$$|x_0| < \frac{r}{K} \quad \text{and} \quad (\beta_1+\beta_2)K/a < 1 \tag{3.17}$$

then there is a unique solution $x(t)$ of (1.16) such that $\forall\ t \geq t_0$,

$$|x(t)| \leq Ke^{-(a-\beta_1 K)(t-t_0)}|x_0| + \frac{K\beta_2}{a-\beta_1 K}r(1-e^{-(a-\beta_1 K)(t-t_0)}) \leq r \tag{3.18}$$

Proof

The complete proof follows standard arguments such as those found in Bellman (1953), Coppel (1965), and Hale (1980). Briefly, referring to Theorem 1.1, let $B = C[t_0,\infty)$ be the Banach space of continuous, bounded $n-$vector functions $y(t)$, defined on $[t_0,\infty)$, and equipped with the norm

$$\|y\| = \sup_{t_0\leq t\leq\infty} |y(t)|$$

The operator T, defined pointwise via

$$(Ty)(t) = F(t,t_0)x_0 + \int_{t_0}^{t} F(t,\tau)[f(\tau,y(\tau))+g(\tau,y(\tau))]d\tau \tag{3.19}$$

is a contraction on the closed subset $R = \{y\in C[t_0,\infty):\|y\|\leq r\}$ of $C[t_0,\infty)$, because for all $\|y_1\|\leq r$ and $\|y_2\|\leq r$

$$|(Ty_1)(t)| \leq Ke^{-a(t-t_0)}|x_0| + K(\beta_1+\beta_2)r\int_{t_0}^{t} e^{-a(t-\tau)}d\tau \leq r \tag{3.20}$$

and

$$\|Ty_1 - Ty_2\| \leq \frac{K(\beta_1 + \beta_2)}{a} \|y_1 - y_2\|$$

By the Contraction Mapping Theorem there exists a unique solution x(t) of (3.19) in R, the fixed point of T. We now prove that (3.17) implies (3.18). From $x(t) = (Tx)(t)$ and (3.19) we have

$$|x(t)| \leq Ke^{-a(t-t_0)}|x_0| + \beta_1 K \int_{t_0}^{t} e^{-a(t-\tau)}|x(\tau)|d\tau + \beta_2 Kr \int_{t_0}^{t} e^{-a(t-\tau)}d\tau, \quad t \geq t_0$$

Using the Bellman-Gronwall Lemma we see that $e^{at}|x(t)| \leq \xi(t)$, $t \geq t_0$, where ξ is the solution of

$$\dot{\xi} = \beta_1 K \xi + \beta_2 Kre^{at} \quad , \quad \xi(t_0) = Ke^{at_0}|x_0|$$

or

$$\xi(t) = e^{\beta_1 K(t-t_0)}\xi(t_0) + e^{at}(1 - e^{-(a-\beta_1 K)(t-t_0)}) \frac{\beta_2 K}{a - \beta_1 K} r$$

which proves (3.18). Under the assumptions, $x = Tx$ is equivalent to (3.13), and hence, the theorem is proved.

□

Remarks: In the bound (3.18), the convergence rate a of the unperturbed system $\dot{x} = A(t)x$ is reduced by $\beta_1 K$. The constant β_1 reflects the size of the perturbation $f(t,x)$, while K increases with the oscillatory character of the unperturbed system. Note also that the allowed range of $|x_0|$ is reduced by the same factor $K \geq 1$. The function $g(t,x)$ differs from $f(t,x)$ in that $g(t,0) \neq 0$, and moreover, in the cases considered here we often have $g(t,x) \equiv g(t)$.

The inequalities in (3.17) can be viewed as defining a *region of attraction* for the set,

$$\lim_{t \to \infty} \sup |x(t)| \leq \frac{K\beta_2}{a - \beta_1 K} r$$

The Total Stability Theorem brings out the important fact that if the unperturbed linear system $\dot{x} = A(t)x$ is exponentially stable, then it is possible to protect against a variety of perturbations, i.e., nonlinearities and additive inputs. The message of this book is reflected exactly in the necessity to produce in $\dot{x} = A(t)x$ an exponential stability.

1.4 THE TUNED SYSTEM AND STRUCTURED ERROR MODELS

The development of the previous sections proceeded formally from the specification of the adaptive system (2.4), through the prescription of some nominal signals e_*, ϕ_*, θ_* to the generation of an error system (3.2), and, via a linearization argument and total stability theorem, to sufficient conditions for boundedness of the solutions of (2.4), or more precisely for closeness of the solutions of (2.4) to the nominal signals. It is clear that the choice of these nominal signals plays a pivotal role in the course of the theory. Since we develop a description (3.2) which consists of a linearizable (or at least Lipschitz continuous) homogeneous part plus additive signals defined by (3.3) we liken the adaptive problem to a nonadaptive problem involving the nominal signals and so denote e_*, ϕ_*, θ_* as "tuned" signals from a possibly fictitious "tuned" system.

There are many feasible alternatives for the choice of tuned signals. In particular, given the requirements of the total stability theorem, one particularly natural choice for e_*, ϕ_*, θ_* is to choose time functions

$$\begin{aligned} e_* &= H_{ew}(\theta_*)w \\ \phi_* &= H_{\phi w}(\theta_*)w \end{aligned} \tag{4.1}$$

so that δ_e, δ_ϕ of (3.3) are zero and then to select θ_* as a constant satisfying some other criterion such as: the value which minimizes the average squared value of e_*, the value which satisfies a constant closed loop nonadaptive control objective, or the value which yields the average value of δ_θ zero.

Such a θ_* would generally be dependent on the signal w, of course, and naturally we would require that $H_{ew}(\theta_*)$ and $H_{\phi w}(\theta_*)$ be stable operators. In effect, a range of possible alternatives is available and different criteria for the choice of e_* could be advanced based on satisfying the requirements of the Total Stability Theorem. In general, the choice of tuned signals affects the solution ball, the Lipschitz constants, and the driving signal bounds so that a trade-off can be contemplated between these competing effects.

While the most convenient choice for tuned signals is as above with θ_* fixed and e_*, ϕ_*, varying with time, when tracking slowly time-varying systems it may be desirable to take θ_* also slowly varying. Equally, it is possible in some instances to choose ϕ_*, say, as the regressor of a model and not composed directly from plant measurements. What is crucial here is not the particular choice of tuned signals since any close choices will produce close conditions for boundedness via the Total Stability Theorem and many of the conceivable algebraic specifications of "best" θ_* will produce similar results.

Another point is that for many adaptive problems the evaluation of tuned signals is unimportant. What is desired by our theory is a collection of sufficient conditions for adaptive systems to have robust bounded solutions and mere existence of tuned solutions satisfying these properties is more important than computation of them. Thus the existence of a fixed parameter setting which produces a stable closed-loop control which is not greatly sensitive to the exact parameter value could give a range of candidate θ_* values and cause to invoke the subsequent definitions of e_*, ϕ_*, to gauge satisfaction of the stability requirements.

In summary then, a plethora of possible tuned signals is conceivable and we do not advocate any particular one. Rather, should the stability requirements be satisfied, then any one of a range of choices is satisfactory. On the other hand, there are many instances where explicit evaluation of tuned signals is inappropriate given the lack of precise plant knowledge which led to the adaptive solution.

1.4.1 Structured Parameter Dependence of the Regressor and Error System

We consider adaptive system structures of the type (2.4) where the explicit $\hat{\theta}$ dependence of $H_{ew}(\hat{\theta})$ and $H_{\phi w}(\hat{\theta})$ is described by

$$\begin{bmatrix} e \\ \phi \end{bmatrix} = \begin{bmatrix} G_{ew} & G_{ev} \\ G_{\phi w} & G_{\phi v} \end{bmatrix} \begin{bmatrix} w \\ v \end{bmatrix} \tag{4.2}$$

where the operators G do not depend on $\hat{\theta}$ and

$$v = \phi^T \hat{\theta} \tag{4.3}$$

Examples of such systems will be given in the next section from which it should be apparent that many standard adaptive systems possess just this structure. As before, we may choose tuned signal values e_*, ϕ_*, θ_* to define

$$v = \phi^T \theta_* - v = \phi^T \theta \tag{4.4}$$

Equation (4.2) takes the form

$$\begin{bmatrix} e \\ \phi \end{bmatrix} = \begin{bmatrix} H_{ew}(\theta_*) & H_{ev}(\theta_*) \\ H_{\phi w}(\theta_*) & H_{\phi v}(\theta_*) \end{bmatrix} \begin{bmatrix} w \\ -v \end{bmatrix} \tag{4.5}$$

where

$$H_{ew}(\theta_*) = G_{ew} - G_{ev}\theta_*^T(I + G_{\phi v}\theta_*^T)^{-1}G_{\phi w}$$

$$H_{ev}(\theta_*) = G_{ev} - G_{ev}\theta_*^T(I + G_{\phi v}\theta_*)^{-1}G_{\phi v} \tag{4.6a}$$

$$= G_{ev}[I + \theta_*^T G_{\phi v}]^{-1} \tag{4.6b}$$

$$H_{\phi w}(\theta_*) = [I+G_{\phi v}\theta_*^T]^{-1}G_{\phi w} \tag{4.6c}$$

$$H_{\phi v}(\theta_*) = [I+G_{\phi v}\theta_*^T]^{-1}G_{\phi v} \tag{4.6d}$$

We remark on several properties of (4.4). Firstly, in this formulation, the signals e and ϕ are the result of applying linear operators to the signals w and v. Further the explicit dependence on the tuned parameter of the operators is displayed in (4.6) and the total error system may be written in incremental form using the definitions (2.5) and (3.3)

$$\begin{aligned} \tilde{e} &= -H_{ev}(\theta_*)v+\delta_e \\ \tilde{\phi} &= -H_{\phi v}(\theta_*)v+\delta_\phi \\ \theta &= \Omega(e,\phi)-\Omega(e_*,\phi_*)+\delta_\theta \\ v &= \phi^T\theta \end{aligned} \tag{4.7}$$

For adaptive systems of the type described by (4.2) and (4.4), Equations (4.7) are the precise form of (3.2). To effect a full linearization as in (3.5) requires that v be replaced by $\phi_*^T\theta$, that the component additive terms be included in (4.7), and that the θ equation be linearized. This will be pursued more fully later but for the moment we take an aside to consider a better structured class of parameter identifiers than the broad class defined by (2.3).

1.4.2 Structures for the Parameter Estimator

The choices of the parameter adaptive algorithm, i.e., the map $\Omega:(\phi,e) \rightarrow \hat{\theta}$, are perhaps unlimited, save that all are usually based on an off-line gradient or Gauss-Newton parameter optimization. Typical algorithms, in continuous-time form, include:

Gradient:

$$\dot{\hat{\theta}} = \frac{\epsilon\phi e}{1+c|\phi|^2}, \quad \epsilon>0 \tag{4.8a}$$

or

$$\dot{\hat{\theta}} = \epsilon\phi e \tag{4.8b}$$

Least Squares:

$$\begin{aligned} \dot{\hat{\theta}} &= \frac{P\phi e}{1 + c\phi^T P\phi} \\ \dot{P} &= -\frac{P\phi\phi^T P}{1+c\phi^T P\phi}, \quad P(0) = P(0)^T > 0 \end{aligned} \tag{4.8c}$$

Whenever $c>0$, the adaptive algorithms are said to be normalized, because the natural continuous-time form would have $c = 0$. Many variants are, of course, possible. For example:

Least Squares with Covariance Resetting:

$$\begin{aligned} \dot{\hat{\theta}} &= \frac{P\phi e}{1+c\phi^T P\phi} \\ \dot{P} &= -\frac{P\phi\phi^T P}{1+c\phi^T P\phi}, \quad P(t_s) = P(0) = P(0)^T > 0 \\ t_s &= \{t \in \mathbb{R}_+ : \min\lambda(P(t)) \leq \lambda_0\} \end{aligned} \tag{4.8d}$$

Gradient with Projection:

$$\dot{\hat{\theta}} = \operatorname*{Proj}_{\hat{\theta}\in\Theta}\left\{\frac{\epsilon\phi e}{1+c|\phi|^2}\right\} \tag{4.8e}$$

where the operator $\operatorname*{Proj}_{\hat{\theta}\in\Theta}$ insures that $\hat{\theta}$ is never moved out of a prescribed region Θ in parameter space. Typically, knowledge of Θ arises from a priori information about the plant.

Throughout the remainder of the text we will simplify our notation and remove the explicit θ_* dependence in H_{ev} and $H_{\phi v}$. Also, for the most part, we will present the analysis of the adaptive error system for a scalar system with a simple gradient algorithm. In this case, our basic adaptive error system is described by

$$
\begin{aligned}
\dot{\theta} &= \epsilon\phi e \\
e &= -H_{ev}v + e_* + \delta_e \\
\phi &= -H_{\phi v}v + \phi_* + \delta_\phi \\
v &= \phi^T\theta
\end{aligned}
\tag{4.9a}
$$

We will often combine these expressions into a shortened form by eliminating e and v. Thus, we have

$$
\begin{aligned}
\dot{\theta} &= -\epsilon\phi H_{ev}\phi^T\theta + \epsilon\phi(e_* + \delta_e) \\
\phi &= \phi_* - H_{\phi v}\phi^T\theta + \delta_\phi
\end{aligned}
\tag{4.9b}
$$

We note that these equations consist of a homogeneous part together with additive driving terms determined by the choice of tuned signals. Their stability properties, therefore, are directly the subject of the Total Stability Theorem. We have two methods for proceeding with this theorem: by direct application to (4.9) or by application to a linearized version of (4.9). It is the latter of these paths which is most informative to follow and so, much of our analysis in Chapters 2 and 3 will be aimed at establishing the exponential asymptotic stability of the differential equation

$$\dot{\theta} = -\epsilon\phi_* H_{ev}(\phi_*\theta) \tag{4.10}$$

Therefore, in Chapters 4 and 5, these results will be invoked together with total stability to demonstrate sufficient conditions for the robust stability of adaptive systems.

It shall usually be the case that H_{ev} and $H_{\phi v}$ are stable finite-dimensional linear systems and, in these instances we may write the error system (4.9) in state-variable form. For example, if

$$
\begin{aligned}
H_{ev}(s) &= d + c^T(sI-A)^{-1}b \\
H_{\phi v}(s) &= D + C^T(sI-F)^{-1}B
\end{aligned}
$$

and we denote their respective states as $x(t)$ and $z(t)$, then (4.9) becomes

$$
\begin{aligned}
\dot{x} &= Ax + b\phi^T\theta \\
\dot{\theta} &= -\epsilon\phi c^T x - \epsilon d\phi\phi^T\theta + \epsilon\phi(e_* + \delta_e) \\
\dot{z} &= Fz + B\phi^T\theta \\
\phi &= -C^T z - D\phi^T\theta + \phi_* + \delta_\phi
\end{aligned}
\tag{4.11}
$$

Thus the adaptive error system may be written completely as a nonlinear state equation which is usually a more familiar and more palatable form than using operators. Moreover, it fits the structure required by the Total Stability Theorem stated for differential equations. The linearization of these state equations is determined by replacing ϕ by ϕ_* as in (4.10) to yield the state-variable form of the linearized adaptive error system

$$
\begin{aligned}
\dot{x} &= Ax + b\phi_*^T\theta \\
\dot{\theta} &= -\epsilon\phi_* c^T x - \epsilon d\phi_*\phi_*^T\theta
\end{aligned}
\tag{4.12}
$$

Since $H_{\phi v}$ is an exponentially stable linear operator we need not consider the ϕ equation in our linearized analysis, e.g., (4.10) is the linearization of (4.9).

1.5 DISCRETE-TIME ADAPTIVE SYSTEMS

The preceding development has taken place using continuous-time. To a large extent this is simply a pedagogical paradigm since many people are more comfortable with analytic theories for differential equations than with equivalent (and often simpler and more direct) theories for difference equations. Nevertheless, discrete-time applications of adaptive control dominate significantly, because of the requirement of computation. It would therefore be remiss of us not to present a complete discrete-time theory as our motivation is the strengthening of adaptive control practice through the development of well directed theory. This distinct but parallel development of the two theories will persist through the book and the emphasis on discrete-time will increase as we turn more to applying the theory from generating it.

The first necessary deviation from the preceding theory of this chapter to include discrete-time is the state-equation statement of the Total Stability Theorem. The only other necessary modification occurs with the specification of the parameter estimation schemes and the subsequent development of state difference equations. Since the two methods are so close we do not dwell long with these adjustments.

Theorem 1.4: Total Stability-Discrete Time

Consider the ordinary difference equation

$$\begin{aligned} x(k+1) &= A(k)x(k)+f(k,x(k))+g(k,x(k)) \\ x(k_0) &= x_0 \in \mathbb{R}^n \end{aligned} \tag{5.1}$$

where $A(k), f(k,x)$ and $g(k,x)$, for each fixed x in the ball $|x| \leq r$, are bounded functions of k, and $\forall |x_1| \leq r$, $\forall |x_2| \leq r$, and $\forall k \geq k_0$:

$$\begin{aligned} &\text{(A1)}\quad f(k,0) = 0 \\ &\text{(A2)}\quad |f(k,x_1)-f(k,x_2)| \leq \beta_1|x_1-x_2| \\ &\text{(A3)}\quad |g(k,x)| \leq \beta_2 r \\ &\text{(A4)}\quad |g(k,x_1)-g(k,x_2)| \leq \beta_2|x_1-x_2| \end{aligned} \tag{5.2}$$

If the unperturbed system

$$x(k+1) = A(k)x(k) \tag{5.3}$$

is exponentially stable, i.e., if for some constants $1>a\geq 0$ and $K\geq 1$, the state transition matrix $F(k_2,k_1)$ of (5.3) satisfies

$$|F(k_2,k_1)| \leq Ka^{k_2-k_1} \quad \forall k_2\geq k_1\geq k_0 \tag{5.4}$$

then

$$|x_0| < \frac{r}{K} \quad \text{and} \quad K(\beta_1+\beta_2)+a<1 \tag{5.5}$$

imply that for $k \geq k_0$

$$|x(k)| \leq K(a+\beta_1 K)^{k-k_0}|x_0| + \frac{K\beta_2 r}{1-(a+\beta_l K)} \tag{5.6}$$

□

As for Theorem 1.3, this result is a direct consequence of the operator-form total stability theorem, Theorem 1.2.

The structure of parameter estimators is probably better known in discrete-time than in the continuous case and the standard forms of gradient algorithms, recursive least squares, and its variants will be presented in detail in Chapter 5. Indeed, we leave until Chapter 5 a full derivation of the error system since this involves careful treatment of normalization, timing, and causality issues, some of which are treated also in Chapter 2. Here it suffices to state that the linearized adaptive error equations equivalent to (4.10) may be written in operator form most simply as the nonstrictly causal difference equation

$$\theta(k+1) \;=\; \theta(k)-\epsilon\phi(k)H_{ev}[\phi^T(k)\theta(k+1)] \tag{5.7}$$

In Chapter 2, a strictly causal state-variable version of (5.7) will be derived but, for the moment, (5.7) best describes the linearized adaptive error equations, highlighting the role played by the respective components.

1.6 CONSOLIDATION

In order to prove robust stability of our adaptive systems by appealing to the Total Stability Theorem, we need first to establish conditions for exponential stability of the linearized error systems

(4.10) and (5.7). This is the topic of Chapters 2 and 3 where restrictions on ϵ, ϕ, and $H_{\epsilon v}$ are introduced and (partially) tied to more direct requirements of signals in adaptive applications.

Chapters 4 and 5 then take particular adaptive algorithms and derive explicit values for $H_{\epsilon v}$, $H_{\phi v}$ and appeal directly to the preceding work for conditions for boundedness of their signals. Specifically, the total stability theorem is appealed to here by: stipulating conditions for the linearized error system to be exponentially stable; stating the system requirements for the additive driving signals to be sufficiently small; describing some possibly suitable choices for tuned signals. Chapter 6 makes this analysis more concrete by considering simulation studies displaying the effects of this theory.

1.7 NOTES AND REFERENCES

The error system notion in adaptive systems is by now classical, establishing the connection between asymptotic stability and convergence, see, e.g., Egardt (1979), Landau (1979), Narendra, Lin, and Valavani (1980), Ljung and Soderstrom (1983), Goodwin and Sin (1984). The operator formulation used in this chapter stems chiefly from the work of Kosut and Friedlander (1982, 1985) and Kosut and Johnson (1984). The application of exponential convergence in the context of robust adaptation is also presented in this latter paper, while more general notions of preservation of exponential stability properties in adaptive error systems are considered by Anderson and Johnstone (1983) and Kosut and Anderson (1986). The Total Stability Theorem also is a classical robustness result available in a form similar to that here in Bellman (1953), Hahn (1967), Coppel (1965), Yoshizawa (1975), Hale (1980), Miller and Michel (1982), and Desoer and Vidyasagar (1975). The operator theoretic notions of linearization, Frechet derivatives, and fixed point theorems may be found in many standard texts, e.g., Luenberger (1969).

Chapter 2

PASSIVITY AND SMALL GAIN ANALYSIS

2.1 INTRODUCTION

Our legacy from the previous chapter and commitment to later ones is to study the behavior, and particularly the stability properties, of the linear system of ordinary differential equations represented by

$$\dot{\theta}(t) = -\epsilon\phi(t)H(s)\{\phi^T(t)\theta(t)\} \tag{1.1}$$

which describe the evolution of the adaptive error system. This system of equations is depicted in Figure 2.1. In discrete-time we equivalently analyze the ordinary difference equation[÷]

$$\theta_{k+1} = \theta_k - \epsilon\phi_k H(z)\{\phi_k^T\theta_{k+1}\} \tag{1.2}$$

As the development of stability theories for (1.2) closely parallels that for (1.1) we shall derive, as often as possible, a unified analysis for both, using concepts from operator theory and specializing where necessary. In particular, the small gain stability theorem will be repeatedly invoked to establish, first, global asymptotic stability of (1.1), (1.2) for all bounded and regulated signals ϕ when H satisfies an SPR (strictly positive real) condition and, thereafter, to establish exponential stability for H SPR together with a restricted class of ϕ, viz., persistently exciting signals.

[÷] In this chapter we use subscript k for discrete time index, and depending on the context, s and z denote either transform variables or the derivative and forward shift operator, respectively.

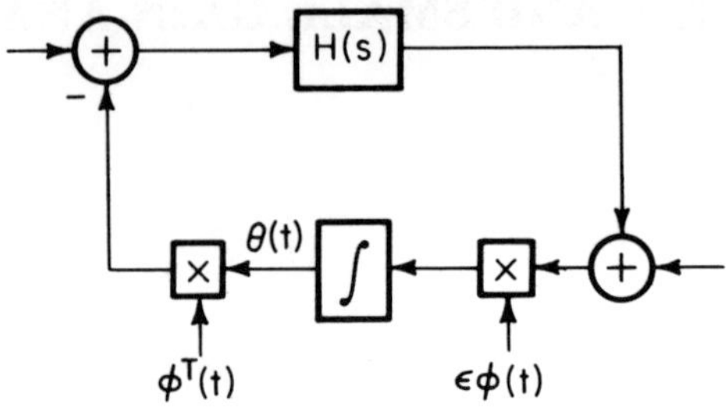

Fig. 2.1 Linear adaptive error system drawn as a feedback loop.

Finally, we consider H non-SPR, which requires a further restriction on the class of ϕ.

Historically, this SPR condition has been a bone of contention in adaptive control for several reasons. First, the appearance in adaptive control theory of such a condition familiar from circuit analysis and synthesis was surprising and served to muddy considerably the waters for newcomers to the field. Its use was in the construction of a Lyapunov function for the complete adaptive scheme using the Popov-Kalman-Yakubovich Lemma. Second, the particular transfer function required to be SPR was not necessarily directly interpretable in terms of plant parameters -- the transfer function only arises in the error system. Third, as was shown in certain celebrated examples due to Egardt, Rohrs, and Ioannou, very slight undermodelling of the plant involved in adaptive control could lead to slight dissatisfaction of the SPR condition and subsequent complete invalidation of the argument used to conclude global stability properties. Adaptive control thus appeared, at least to some researchers, inherently non-robust.

In this and subsequent chapters, we shall attempt to dispel some of these misapprehensions by carefully developing sufficient conditions for robust stability of (1.1), (1.2) together with explanations of the underlying intuitions. We begin by treating the necessary mathematical formalisms and presenting the small gain theorem.

2.2 MATHEMATICAL PRELIMINARIES AND THE SMALL GAIN THEOREM

This section addresses the formalities of setting out the mathematical scenery against which our development takes place. Our notation and conventions follow closely those of Desoer and Vidyasagar (1975) and Zames (1966) and our aim here is to set up quickly the backdrop for a statement of the small gain theorem.

We consider here the linear space F of real valued functions $f : \mathbf{T} \to \mathbb{R}^n$ where $\mathbf{T} = \mathbb{R} = \{t|-\infty < t < \infty\}$ or $\mathbb{R}_+ = \{t|0 \leq t < \infty\}$ for continuous-time and $\mathbf{T} = \mathbf{Z} = \{\ldots, -1,0,1, \ldots\}$ or $\mathbf{Z}_+ = \{0,1,2, \ldots\}$ for discrete-time. We denote by $|\cdot|$ any suitable norm on $\mathbb{R}^n$ and consider a norm $\|\cdot\|$ on F. Typically for continuous-time we consider the L_p norms

$$\|f\|_p = \left(\int_{t \in \mathbf{T}} |f(t)|^p dt \right)^{1/p} \tag{2.1}$$

and, for discrete-time, the l_p norms

$$\|f\|_p = \left(\sum_{t \in \mathbf{T}} |f(t)|^p \right)^{1/p} \tag{2.2}$$

where $p \in (0,\infty)$ or the L_∞ (resp. l_∞) norm

$$\|f\|_\infty = \sup_{t \in \mathbf{T}} |f(t)| \tag{2.3}$$

Within the space F we shall restrict our attention to a normed linear subspace L,

$$L = \{f: \mathbf{T} \to \mathbb{R}^n | \, \|f\| < \infty\} \tag{2.4}$$

or to an extended normed linear subspace L_e which admits certain functions of unbounded norm

$$L_e = \{f: \mathbf{T} \to \mathrm{I\!R}^n | \forall\ 0 < T \in \mathbf{T}\ ,\ \|f_T\| < \infty\} \tag{2.5}$$

where

$$f_T(t) = \begin{cases} f(t) & \text{for} \quad |t| \le T \\ 0 & \text{for} \quad |t| > T \end{cases} \tag{2.6}$$

is the truncation (or projection) of f. The function spaces L_p, L_{pe}, l_p, and l_{pe} will be those L or L_e produced by considering the respective norms from (2.1), (2.2), or (2.3).

In this formalism, given their initial conditions, systems can be described by operators mapping an input function space L or L_e into an output function space L or L_e. It is therefore of interest to study more specifically input-output properties of systems. This general framework admits the unified treatment of linear, non-linear, time-varying, non-causal, discrete-time, etc., systems which, in our context of adaptive control, will permit the development of stability theorems applicable to broad classes of systems capable of encompassing a range of uncertainty and thus leading to robustness.

We shall concentrate on the stability properties of general feedback interconnections of the type depicted in Figure 2.2, where $H_1, H_2: L_e \to L_e$; cf. Figure 2.1.

Definition: Given an operator $H: L_e \to L_e$, suppose that real numbers $\bar{\gamma}$, $\bar{\beta}$ exist such that

$$\|(Hf)_T\| \le \bar{\gamma}\|f_T\| + \bar{\beta} \tag{2.7}$$

for all $f \in L_e$, $T \in \mathbf{T}$. Then we define the gain, $\gamma(H)$, of operator H as the infimum of all $\bar{\gamma}$ such that there exists a $\bar{\beta}$ for which (2.7) holds.

In intuitive terms $\gamma(H)$ describes the increase in norm of an output function from the input function to H. For causal H the truncation may be dropped and only $f \in L$ need be used to define the gain. We should remark here that the particular value of gain, or

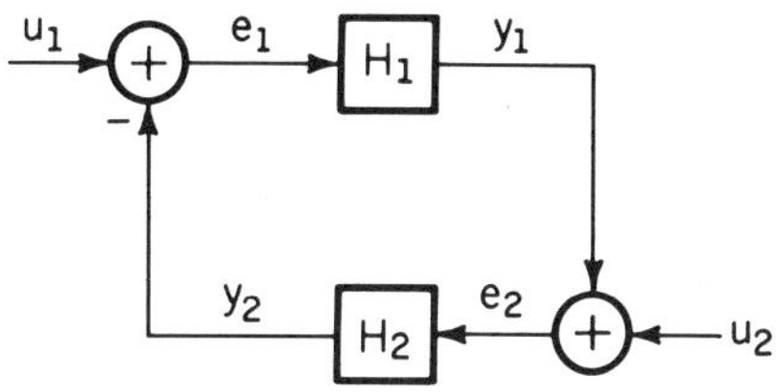

Fig. 2.2 Feedback interconnection of systems.

even its existence, for an operator may depend critically upon the particular norm chosen -- for example, an ideal relay has zero L_∞ gain but infinite L_p gain for $p < \infty$!

In our subsequent analysis we shall repeatedly utilize certain operator gains and so will need the following results which, generally, are easily derived from the definition.

Lemma 2.1:

(i) The linear operator $y(t) = h(t)u(t)$ for $h(t) \in L_\infty$ has L_p (resp. l_p) gain $\gamma = \|h(\cdot)\|_\infty$ from u to y.

(ii) The convolution operator $y(t) = (h * u)(t)$ has $L_p(l_p)$ gain bounded above by $\|h(\cdot)\|_1$ for all $p \in [1,\infty]$.

(iii) The linear nonanticipative operator with impulse response, or kernel, $F(t,\tau)$ satisfying, for all $t > \tau$,

$$F(t,\tau) \leq Ke^{-a(t-\tau)} \quad , \quad a > 0 \, , K \geq 0$$

has L_p gain bounded above by K/a for all $p \in [1,\infty)$.

(iv) The stable, causal, linear, time-invariant operator with Laplace transform representation $H(s)$ and impulse response $H(t)$ has L_∞ gain

$$\gamma_\infty(H) = \int_0^\infty |h(\tau)| d\tau$$

and L_2 gain

$$\gamma_2(H) = \max_{\omega \in \mathbb{R}} |H(j\omega)| \leq \gamma_\infty(H)$$

For discrete-time systems:

(iii′) The linear nonanticipative operator with impulse response $F(t,\tau)$ satisfying, for all $t > \tau$,

$$|F(t,\tau)| \leq Ka^{t-\tau} \ , \ |a| < 1 \ , \ K \geq 0$$

has l_p gain bounded above by $K(1-a)$.

(iv′) The stable, linear, causal, time-invariant operator with $z-$transform representation $H(z)$ and impulse response $h(t)$ has l_∞ gain

$$\gamma_\infty(H) = \sum_{t=0}^\infty |h(t)|$$

and l_2 gain

$$\gamma_2(H) = \sup_{\theta \in [0,2\pi)} |H(e^{j\theta})| \leq \gamma_\infty(H)$$

Returning to the feedback interconnection of Figure 2.2, we have

Theorem 2.1: The Small Gain Theorem

Consider the system shown in Figure 2.2. Let H_1, H_2: $L_e \rightarrow L_e$. Suppose that $u_1, u_2 \in L_e$ implies the existence of $e_1, e_2 \in L_e$ and, further, that H_1 and H_2 have finite gains γ_1, γ_2 and associated constants β_1, β_2 from (2.7). If, under these conditions,

$$\gamma_1\gamma_2 < 1 \tag{2.8}$$

then

(i) If $u_1, u_2 \in L_e$,

$$\|e_{1T}\| \leq (1-\gamma_1\gamma_2)^{-1}(\|u_{1T}\| + \gamma_2\|u_{2T}\| + \beta_2+\gamma_2\gamma_1)$$

$$\|e_{2T}\| \leq (1-\gamma_1\gamma_2)^{-1}(\|u_{2T}\| + \gamma_2\|u_{1T}\| + \beta_1+\gamma_1\gamma_2) \qquad (2.9)$$

for all $T \in \mathbf{T}$.

(ii) If $u_1, u_2 \in L$, then $e_1, e_2, y_1, y_2 \in L$ and the norms of e_1 and e_2 are bounded by the right-hand side above with nontruncated functions.

□

The proof of the small gain theorem follows from (2.9) by substitution. Its statement is that the feedback interconnection of finite gain systems as in Figure 2.2 will be bounded input (u_1,u_2) bounded output $(e_1,e_2$ or $y_1,y_2)$ stable provided the product of the gains is less than one. For linear, time-invariant systems and the L_2 norm, this condition has a direct interpretation as a magnitude-only formulation of the Nyquist stability criterion. As stated above, there is a requirement of existence of solutions $e_1,e_2 \in L_e$ when $u_1,u_2 \in L_e$. This condition may be removed for non-anticipative (causal) H_1,H_2 which satisfy a global Lipschitz continuity condition. This is a strong condition. Throughout our analysis in this book the more specific existence questions will be addressed for each particular problem by appealing to the Total Stability Theorem.

With the small gain theorem by our side we are now prepared to address the questions of stability of the linear adaptive error systems (1.1), (1.2). We begin by proving the global stability of (1.1) for arbitrary signals $\phi \in L_\infty$ when H satisfies a strictly positive real (SPR) condition. For ϕ in a reduced subset of L_∞ -- the set of persistently exciting signals -- (1.1) will be shown to be exponentially stable for SPR H.

In order to maintain exponential stability, when the SPR condition is relaxed it is necessary to restrict the class of ϕ, and in particular, when ϵ is small we shall investigate such a class of functions and its interpretation. Our analysis will concentrate on the continuous-time error system (1.1) initially and we shall specialize to the discrete-time error system (1.2) later so as not to clutter the presentation.

2.3 EXPONENTIAL STABILITY, THE SPR CONDITION, PASSIVITY

We shall consider here conditions on $\phi(t)$ and $H(s)$ which ensure global stability properties of (1.1) and especially global exponential stability. We presume that $H(s)$ is a rational transfer function which is proper, i.e., $|H(\infty)| < \infty$, with minimal state variable realization A,b,c,d, i.e.

$$H(s) = d + c^T(sI-A)^{-1}b \tag{3.1}$$

Then (1.1) may be more fully written as a linear time-varying ordinary differential equation involving the state x of $H(s)$ in addition to θ:

$$\begin{bmatrix} \dot{x}(t) \\ \dot{\theta}(t) \end{bmatrix} = \begin{bmatrix} A & b\phi^T(t) \\ -\epsilon\phi(t)c^T & -\epsilon d\phi(t)\phi^T(t) \end{bmatrix} \begin{bmatrix} x(t) \\ \theta(t) \end{bmatrix} \tag{3.2}$$

2.3.1 A More General Framework

To proceed from here with (3.2) we consider the more general problem of exponential stability of the linear, time-varying state equation

$$\dot{x} = F(t)x \tag{3.3}$$

for general matrices $F(t)$. We have the following result.

Lemma 2.2:

Consider the differential equation (3.3) where the matrix function $F(\cdot)$ is bounded and locally integrable. Suppose that there exists a positive definite constant symmetric matrix P such that

$$PF(t) + F^T(t)P = -N(t)N^T(t) \tag{3.4}$$

for some matrix function $N(\cdot)$ and all t. Then (3.3) is stable in the sense of Lyapunov.

If, further, the pair $[F(t), N(t)]$ is uniformly completely observable, that is, writing $\Phi(t,\tau)$ as the transition function of (3.3), there exist finite constants $T > 0$, $\beta > \alpha > 0$ such that for all t

$$\beta I \geq \int_t^{t+T} \Phi^T(\tau,t)N(\tau)N^T(\tau)\Phi(\tau,t)d\tau \geq \alpha I \tag{3.5}$$

then (3.3) is exponentially stable.

Proof

Write the Lyapunov function $V(t) = x^T(t)Px(t)$. Then, from (3.4) we have $\dot{V}(t) = -x^T(t)N(t)N^T(t)x(t) \leq 0$, and stability follows.

To achieve exponential stability, consider

$$\begin{aligned}
V(t+T)-V(t) &= \int_t^{t+T} \dot{V}(\tau)d\tau \\
&= -\int_t^{t+T} x^T(\tau)N(\tau)N^T(\tau)x(\tau)d\tau \\
&= -\int_t^{t+T} x^T(t)\Phi^T(\tau,t)N(\tau)N^T(\tau)\Phi(\tau,t)x(t)d\tau \\
&= -x^T(t)\{\int_t^{t+T} \Phi^T(\tau,t)N(\tau)N^T(\tau)\Phi(\tau,t)d\tau\}x(t) \\
&\leq -\frac{\alpha}{\lambda_{max}(P)} V(t)
\end{aligned}$$

This last inequality follows from the Lemma conditions. We know also that $V(t)$ is non-increasing so that $V(t+\tau) \leq V(t)$ for all $\tau \geq 0$ and $t \geq 0$. Hence

$$V(t+\tau) \leq \left(1 - \frac{\alpha}{\lambda_{max}P}\right)^{\left(\frac{\tau}{t} - 1\right)} V(t)$$

for all $\tau \geq 0$ and $t \geq 0$. Thus, $V(t)$ is exponentially decreasing and, as P is a constant positive definite matrix, this implies that $x(t)$ converges to zero exponentially fast.

□

Before applying this Lemma directly to (1.1) via (3.2) we state one more simplifying result which will facilitate the direct verification of exponential stability of (3.2).

Lemma 2.3:

The pair $[F(t),N(t)]$ will be uniformly completely observable, i.e., satisfy (3.5), if and only if the pair $[F(t)-K(t)N^T(t),N(t)]$, with $K(t)$ bounded and locally integrable, is uniformly completely observable.

Proof

By inspection of Figure 2.3 observability properties from measurements of $u(t)$ and $y(t)$ are the same as those from $v(t)$ and $y(t)$ since $u(t) = v(t)-K(t)y(t)$ is known.

□

We are now in a position to apply these general results to our particular variant (3.2) of (3.3). But firstly, we present a simple example.

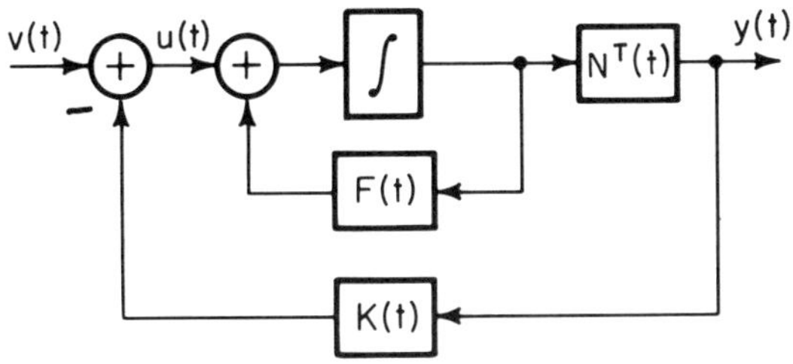

Fig. 2.3 Bounded output feedback in state variable form.

2.3.2 A Simple Example

To fix our thinking concerning conditions for the stability of the error system (1.1), consider the simplest case where $H(s)$ is a non-dynamic constant h. Equation (1.1) then has the form

$$\dot{\theta}(t) = -\epsilon h \phi(t)\phi^T(t)\theta(t) \tag{3.6}$$

Using the results from the previous subsection we have

Theorem 2.2:

For $\phi(t)$ bounded and locally integrable, (3.6) is globally exponentially stable for any $\epsilon > 0$ if

(i) $h > 0$ (3.7)

(ii) There exist positive constants T, α, β such that for all t

$$\infty > \beta I \geq \int_t^{t+T} \phi(\tau)\phi^T(\tau)d\tau \geq \alpha I > 0 \tag{3.8}$$

Proof

Take $P = I$ in Lemma 2.2, noting that here $F(t) = -\epsilon h \phi(t)\phi^T(t)$ to yield $N(t) = \gamma\phi(t)$ where $\gamma = (2\epsilon h)^{1/2}$. Note this requires $h > 0$. To establish exponential stability we need to consider the uniform complete observability of $[F(t), N(t)]$ or, from Lemma 2.3, that of $[F(t) - K(t)N^T(t), N(t)]$ for bounded $K(t)$. Take $K(t) = -\frac{\gamma}{\epsilon h}\phi(t)$ to yield $F - KN^T \equiv 0$ and hence $\Phi(t,\tau) = I$ in (3.5).

□

We offer some immediate comments on the substance of this result. From our notions of stability for scalar ordinary differential equations like $\dot{x} = -a(t)x$ it is not surprising that condition (i) (3.7) arises since the dyadic term $\phi(t)\phi^T(t)$ is nonnegative definite by construction. This connection to a simple scalar equation may be further extended to argue that the uniform positivity of the function $a(t)$ above dictates the exponential rate. Condition (ii) (3.8), known as *persistence of excitation,* clearly plays a similar role here. Persistence of excitation will be a recurring theme throughout this book and primarily ensures that $\theta(t) = 0$ is the only solution of $\dot{V} = 0$, i.e., it is the only equilibrium solution, and through its uniformity in t that convergence to this equilibrium is uniform. For linear systems, this means exponential stability.

We now move on from this simple example to more general $H(s)$.

2.3.3 The SPR Condition

In the light of conditions (3.7), (3.8), we may now reappraise (1.1) for more general $H(s)$. Studying (1.1) we see that $\dot{\theta}(t)$ is a vector parallel to $\phi(t)$ with a scalar multiplier of $-\epsilon H(s)\{\phi^T(t)\theta(t)\}$. To preserve the stability properties of the previous subsection it is reasonable to require that the sign and magnitude properties of $H(s)\{\phi^T(t)\theta(t)\}$ be the same as those of $\phi^T(t)\theta(t)$.

If without qualifying any other property of $\phi^T(t)\theta(t)$ we make just the restriction that the sign of $\phi^T(t)\theta(t)$ is preserved at every time instant after passing through H, then we limit the scope of admissible $H(s)$. Instead:

Lemma 2.4:

Suppose that $H(s)$ has no poles in $\mathrm{Re}(s) > 0$ and satisfies

(i) The poles of $H(s)$ which are on the $j\omega$-axis are simple and such that the associated residue is non-negative,

(ii) For any real ω for which $j\omega$ is not a pole of $H(s)$,

$$\mathrm{Re}[H(j\omega)] \geq 0 \tag{3.9}$$

Then there is a constant γ depending on initial conditions in H such that for any function $x \in L_{2e}$

$$\int_0^t x(\tau)(Hx)(\tau)d\tau \geq -\gamma \tag{3.10}$$

for all t.

□

This result may be established by Parseval's theorem. For our purposes, the interpretation of (3.10) is that $H(s)$ is a sign-preserving operator, at least in an integral sense, over any nonzero time interval. The class of linear time invariant systems satisfying (3.9) is familiar from circuit theory when the integral (3.10) is interpreted as an energy function.

Definition: The time-invariant linear system with transfer function $H(s)$ which satisfies the former assumptions of the Lemma is called *positive real*. The system is *strictly positive real (SPR) if* $H(s-\mu)$ *is positive real for some* $\mu > 0$.

The class of SPR systems arises in several seemingly disparate disciplines but is most familiar from circuit theory where passive, linear, time-invariant network admittance transfer functions are SPR. The notion of passivity can be extended to time-varying and nonlinear operators. Perhaps the most frequent perplexity of adaptive systems has been the occurrence of the SPR condition seemingly out of context. In terms of its sign preservation properties this now should be less mysterious. Another result connected with SPR transfer functions is the Popov-Kalman-Yakubovich Positive Real Lemma.

Lemma 2.5: The Positive Real Lemma

Let $H(\cdot)$ be a real rational function of a complex variable s, with $|H(\infty)| < \infty$. Let A,b,c,d be a minimal realization of $H(s)$, i.e.,

$$H(s) = d + c^T(sI-A)^{-1}b \tag{3.11}$$

Then $H(s)$ is strictly positive real if and only if there exist real matrices P and L and constants w and γ with P positive definite and $\gamma^2=2\mu$ above such that

$$\begin{aligned} PA + A^TP &= -L\,L^T - \gamma^2 P \\ Pb &= c - wL \\ w^2 &= 2d \end{aligned} \tag{3.12}$$

□

Again, this result has a heritage in network synthesis but, for our purposes we shall use it to define a new state-variable realization for SPR $H(s)$.

Lemma 2.6:

Let $H(s)$ be an SPR rational transfer function with $H(\infty) < \infty$. Then $H(s)$ has a state variable realization A,b,c,d where

$$A + A^T = -QQ^T - \gamma^2 I \tag{3.13}$$

$$b = c - wQ \tag{3.14}$$

$$w^2 = 2d \tag{3.15}$$

for some non-zero matrix Q and non-zero constant γ.

Proof

Multiply the equations of the Positive Real Lemma on the left and on the right by $P^{1/2}$, a symmetric square root of P, to define a new state variable realization $\bar{A} = P^{1/2}AP^{-1/2}$, $\bar{b} = P^{1/2}b$, $\bar{c} = cP^{-1/2}$, $\bar{d} = d$ which satisfies (3.13), (3.14) with $Q = P^{-1/2}L$.

□

Having presented and developed the SPR condition for a transfer function we now press on to establish rigorously its role in the global exponential stability of (1.1).

2.3.4 Exponential Stability with the SPR Condition

Using Lemma 2.6, we may write (1.1) or (3.2) in the form

$$\begin{bmatrix} \dot{x}(t) \\ \dot{\theta}(t) \end{bmatrix} = \begin{bmatrix} A & \epsilon^{1/2} b \phi^T(t) \\ -\epsilon^{1/2}\phi(t)b^T - \epsilon^{1/2} w \phi(t) Q^T & -\epsilon d \phi(t)\phi^T(t) \end{bmatrix} \begin{bmatrix} x(t) \\ \theta(t) \end{bmatrix} \tag{3.16}$$

without loss of generality. From here we may invoke the stability techniques of Lemma 2.2 and Lemma 2.3. The co-ordinate basis change to allow the state variable description (3.16) admits the choice of $P = I$ for our Lyapunov function although the more usual approach is to use the P of the Popov-Kalman-Yakubovich Lemma directly with its respective state variable realization.

Theorem 2.3:

Consider the ordinary differential equation (1.1)

$$\dot{\theta}(t) = -\epsilon\phi(t)H(s)\{\phi^T(t)\theta(t)\}$$

where $H(s)$ is a strictly positive real rational proper transfer function and $\phi(t)$ is bounded and locally integrable. Then this differential equation is stable and, further, is exponentially stable (i.e., $\theta(t)$ and the state of $H(s)$ decay exponentially fast to zero) if either:

(i) $0 < H(\infty)$ and there exist positive constants T, α and β such that, for all t,

$$\infty > \beta I \geq \int_t^{t+T} \phi(\tau)\phi^T(\tau)d\tau \geq \alpha I > 0 \tag{3.17}$$

or

(ii) $H(\infty) = 0$ and there exist positive constants T', α' and β' such that, for all t

$$\infty > \beta' I \geq \int_t^{t+T'} \int_t^{\tau} \phi(\sigma)d\sigma \int_t^{\tau} \phi^T(\sigma)d\sigma d\tau \geq \alpha' I > 0 \tag{3.18}$$

Further, if $\phi(t)$ has bounded derivative for all t except possibly for a countable number of points of fixed minimum separation, then (3.17) is equivalent to (3.18).

Proof

Since $H(s)$ is rational and SPR we may take the state-variable realization for H given in Lemma 2.6 and, hence, take our differential equation in the form of (3.16). We now use Lemma 2.2

with $P = I$ and consider, writing $F(t)$ for the matrix in (3.16) and $w^2 = 2d$ from (3.15), and absorbing the factor $\epsilon^{1/2}$ into $\phi(t)$,

$$F(t) + F^T(t) = \begin{bmatrix} -QQ^T-\gamma^2 I & -wQ\phi^T(t) \\ -w\phi(t)Q^T & -w^2\phi(t)\phi^T(t) \end{bmatrix}$$

$$= -\begin{bmatrix} Q & \gamma I \\ w\phi(t) & 0 \end{bmatrix}\begin{bmatrix} Q^T & w\phi^T(t) \\ \gamma I & 0 \end{bmatrix}$$

We have now identified the matrix $N(t)$ of (3.4) and the stability of the differential equation is immediate. To achieve exponential stability we need to consider the uniform complete observability of $[F(t),N(t)]$. Appealing to Lemma 2.3 we will examine the observability of $[F(t)-K(t)N^T(t),N(t)]$. Let

$$K(t) = \begin{bmatrix} K_1 & K_2 \\ K_3 & K_4 \end{bmatrix}$$

with partitions commensurate with those of $F(t)$. Then

$$F(t)-K(t)N^T(t) =$$
$$\begin{bmatrix} A-K_1Q^T-\gamma K_2 I & b\phi^T(t)-wK_1\phi^T(t) \\ -\phi(t)b^T-w\phi(t)Q^T-K_3Q^T-\gamma K_4 I & -d\phi(t)\phi^T(t)-wK_3\phi^T(t) \end{bmatrix}$$

We now consider two cases:

(i) $0 < H(\infty) < \infty$. Thus $d > 0$, $w \neq 0$ and we choose

$$K(t) = \begin{bmatrix} w^{-1}b & \gamma^{-1}(A-w^{-1}bQ^T)I \\ -\frac{w}{2}\phi(t) & -\gamma^{-1}(\phi(t)b^T + \frac{w}{2}\phi(t)Q^T)I \end{bmatrix}$$

to yield

$$F(t) - K(t)N^T(t) = \begin{bmatrix} 0 & 0 \\ 0 & 0 \end{bmatrix}$$

The transition matrix $\Phi(t,\tau)$ for this $F-KN^T$ is I for all τ,t and so the observability gramian for this $[F-KN^T,N]$ is

$$\int_t^{t+T} \begin{bmatrix} QQ^T+\gamma^2 I & wQ\phi^T(\sigma) \\ w\phi(\sigma)Q^T & w^2\phi(\sigma)\phi^T(\sigma) \end{bmatrix} d\sigma$$

Condition (3.17) is *necessary and sufficient* for this observability gramian to satisfy (3.5).

(ii) $H(\infty) = 0$. Thus $d = w = 0$ and we select

$$K(t) = \begin{bmatrix} 0 & \gamma^{-1}AI \\ 0 & -\gamma^{-1}\phi(t)b^T I \end{bmatrix}$$

producing

$$F(t)-K(t)N^T(t) = \begin{bmatrix} 0 & b\phi^T(t) \\ 0 & 0 \end{bmatrix}$$

The transition matrix is given by

$$\Phi(\tau,t) = \begin{pmatrix} I & \int_t^\tau b\phi^T(\sigma)d\sigma \\ 0 & I \end{pmatrix}$$

and the observability gramian of $[F-KN^T,N]$ is

$$O(t+T,t) =$$

$$\int_t^{t+T} \begin{pmatrix} QQ^T+\gamma^2 I & (QQ^T+\gamma^2 I)b\int_t^\tau \phi^T(\sigma)d\sigma \\ \int_t^\tau \phi(\sigma)d\sigma b^T(QQ^T+\gamma^2 I) & \int_t^T \phi(\sigma)d\sigma b^T(QQ^T+\gamma^2 I)b\int_t^\tau \phi^T(\sigma)d\sigma \end{pmatrix} d\tau$$

$$= \begin{pmatrix} TD & Db\int_t^{t+T} W^T(\tau)d\tau \\ \int_t^{t+T} W(\tau)d\tau b^T D & b^T Db\int_t^{t+T} W(\tau)W^T(\tau)d\tau \end{pmatrix}$$

with $D = QQ^T + \gamma^2 I$ and $W(\tau) = \int_t^\tau \phi(\sigma) d\sigma$.

We now show that uniform non-singularity of $O(t+T,t)$ is equivalent to uniform non-singularity of

$$R(t+T,t) = \int_t^{t+T} W(\tau) W^T(\tau) d\tau$$

That singularity of $R(t+T,t)$ implies singularity of $O(t+T,t)$ is immediate because if there exists a non-zero vector x_2 satisfying $R(t+T,t)x_2 = 0$, then also $\int_t^{t+T} W^T(\tau) d\tau x_2 = 0$ and so $x = [0 \;\; x_2^T]^T$ is a nonzero null vector of $O(t+T,t)$. To establish the reverse implication, let $x = [x_1^T \;\; x_2^T]^T$ where x_1 and x_2 are any two suitably dimensioned unit vectors. Further, denote by $D^{1/2}$ a square root of D. We have

$$[x_1^T \; x_2^T] O(t+T,t) \begin{bmatrix} x_1 \\ x_2 \end{bmatrix} = \int_t^{t+T} \| D^{1/2}(x_1 + bW^T(\tau)x_2 \|^2 d\tau$$

$$= S(t,t+T)$$

Now notice that the integral form of $W(\tau)$ and the Schwarz inequality imply for all $\Delta \epsilon [t,t+T]$ that

$$1/2 | \| D^{1/2}(x_1 + bW^T(\Delta)x_2) \|^2 - x_1^T D x_1 |$$

$$= | \int_t^\Delta x_2^T \dot{W}(\tau) b^T D(x_1 + bW^T(\tau)x_2) d\tau |$$

$$\leq \left(\int_t^\Delta \| D^{1/2} b \dot{W}^T(\tau) x_2 \|^2 d\tau \right)^{1/2} \left(\int_t^\Delta \| D^{1/2}(x_1 + bW^T(\tau)x_2) \|^2 d\tau \right)^{1/2}$$

$$\leq (b^T D b)^{1/2} T^{1/2} \sup_{\tau \in [t,t+T]} \| \phi(\tau) \| S^{1/2}(t,t+T)$$

Integrating this inequality for Δ from t to $t+T$ we get

$$T x_1^T D x_1 \leq S(t,t+T) + 2T^{3/2}(b^T D b)^{1/2}$$

$$\sup_{\tau \in [t,t+\tau]} \|\phi(\tau)\| S^{1/2}(t,t+T)$$

while, on the other hand, the triangle inequality yields also that

$$\int_t^{t+T} \|D^{1/2}bW^T(\tau)x_2\|^2 d\tau \leq \int_t^{t+T} x_1^T D x_1 d\tau + S(t,t+T)$$

Hence

$$x_2^T \int_t^{t+T} W(\tau)W^T(\tau)d\tau x_2 \leq \frac{2}{b^T Db} \{S(t,t+T) + kS^{1/2}(t,t+T)\}$$

with k representing the constant above.

This argument shows that uniform positive definiteness of $R(t+T,t)$ implies uniform positive definiteness of $O(t+T,t)$. The exponential stability of (1.1) follows.

We do not offer a formal proof of the final statement of the theorem but illustrate it by simply noting that if $\phi(t)$ has increasingly rapid switching, as for example do $\phi(t) = \sin(t^2)\phi_0$ or $\phi(t) = \text{sgn}\cos(e^t)\phi_0$, then there exists no fixed T for which (3.17) and (3.18) are equivalent because of the effect of integration of the signals in spite of their great variation over the interval of integration.

□

Theorem 2.3 is of fundamental importance in our development of adaptive systems. Recall that (1.1) describes the evolution of parameter estimation error in our adaptive system. Exponential stability of (1.1) is equated with exponential convergence of the adaptive estimator. The SPR condition on $H(s)$ pertains to properties of the estimation algorithm being used and to the actual plant involved. The next section will be devoted to analysis of this condition and of its relaxation. The signal properties of (i) and (ii) are of practical, theoretical and even philosophical importance. We shall refer to (3.17) as the *persistence of excitation* condition since it constrains the vector-signal $\phi(t)$ to be spanning in every direction by a minimum component, uniformly in time. The implications of and requirements for persistence of excitation will be discussed more fully in Section 2.5 but next we move on to consider relaxation of the SPR condition by further limiting the class of signals.

Before moving on, we conclude this section with a simple result which allows a broad quantification of the convergence rate of (1.1) when the adaptation gain, ϵ, is small. Theorem 2.3 is a very strong stability result, being applicable for broad classes of ϕ and it will prove necessary later to quantify the effects of ϵ on the convergence rate obtained for SPR $H(s)$ with given $\phi(t)$ which is persistently exciting. We have:

Corollary 2.1:

Suppose that $H(s)$ is SPR and that $\phi(t)$ satisfies the persistence of excitation conditions of Theorem 2.3. Then the degree of stability, i.e., the rate of exponential stability, of (1.1) is proportional to ϵ for ϵ sufficiently small.

Moreover, the magnitude of the state of (3.2) decays as

$$\left\| \begin{bmatrix} x(t+\tau) \\ \theta(t+\tau) \end{bmatrix} \right\| \leq me^{-\lambda\tau} \left\| \begin{bmatrix} x(t) \\ \theta(t) \end{bmatrix} \right\|$$

for all $t, \tau > 0$ where, for small ϵ, there exist positive constants k_λ and k_m such that

$$\lambda = k_\lambda \epsilon + O(\epsilon^2)$$

and

$$m = k_m + O(\epsilon)$$

Proof

The first part of this result may be proven by quantifying the minimum eigenvalue of the observability gramians of the pair $[F(t),N(t)]$ in the proof of Theorem 2.3 -- noting the absorption of

$\epsilon^{1/2}$ into $\phi(t)$ in this proof. This minimum eigenvalue scales linearly with ϵ for ϵ small. This somewhat tortuous path to establishing this result is superceded by the more elegant averaging methods of Chapter 3 and so we leave the full proof of the corollary as a (post-Chapter 3) exercise for the reader.

The second part stems from the observation that the quadratic Lyapunov function $z^T P z$ implies from its decrescent property that

$$z^T(t+\tau)Pz(t+\tau) \leq z^T(t)Pz(t)$$

and hence for any $\tau > 0$ and all t

$$|z(t+\tau)|^2 \leq \frac{\lambda_{\max}(P)}{\lambda_{\min}(P)} |z(t)|^2$$

Combining this with the result of Theorem 2.3

$$|z(t+T)|^2 \leq \exp(-2\lambda T)|z(t)|^2$$

for T of the persistence condition, we have

$$|z(t+\tau)| \leq \left(\frac{\lambda_{\max}(P)}{\lambda_{\min}(P)} \right)^{1/2} e^{\lambda T} \cdot e^{-\lambda\tau}|z(t)|$$

for all $t, \tau > 0$. Identifying here that $\lambda_{\max}(P) = \lambda_{\min}(P) = 1$ and $\lambda = k_\lambda \epsilon + O(\epsilon^2)$ yields the quantification of m.

□

2.3.5 A Brief Respite and Some Commentary

Following the technical rigors of the section so far and before plunging headlong into the next section, it is an opportune juncture to take stock of the results which we have generated and to discuss and explain some of their requirements and ramifications.

Recall that our task is to study the stability properties of the linearized adaptive error system equations (1.1)

$$\dot{\theta}(t) = -\epsilon\phi(t)H(s)\{\phi^T(t)\theta(t)\}$$

where: $\theta(t)$ represents the parameter estimate deviation from its nominal value; ϵ is the gain (or step size) of the adaptation algorithm; $\phi(t)$ is the regression vector composed of plant inputs and outputs; and $H(s)$ is a transfer function determined by the underlying plant and the adaptation algorithm. The major results developed so far are stated in Theorem 2.3 and Corollary 2.1, but it is also helpful to keep in mind the simple preliminary example of Subsection 2.3.2, whose similarity in turn, to the scalar ordinary differential equation $\dot{x} = -\epsilon a(t)x$ is also worth noting.

The conditions of Theorem 2.3 for exponential stability are first that $H(s)$ should be SPR and then, second, that $\phi(t)$ should be persistently exciting in accordance with (3.17) or (3.18). Provided $\dot{\phi}$ is bounded, the former of these conditions is sufficient for stability of (1.1). As argued in Subsection 2.3.3 the requirement of SPR for $H(s)$ is interpretable in terms of sign preservation of the input signal, at least in the integral sense necessary for stability. The transition from stability to exponential stability requires a uniformity of magnitude of $\dot{\theta}$ in addition to sign preservation properties. The interpretation of the lower bound in (3.17) is precisely that the regression vectors should have this uniformity of magnitude while the modified condition (3.18) embodies the requirement that the interposing transfer function $H(s)$ should not attenuate the signal $\phi^T(t)\theta(t)$ progressively more with time by, for example, having the dominant frequency content of $\phi(t)$ occurring at increasingly large frequencies when $H(\infty) = 0$.

When $H(s)$ is SPR, stability of (1.1) follows without conditions on the regression vector $\phi(t)$. However, for non-SPR $H(s)$ it is possible to make (1.1) unstable by choice of $\phi(t)$ -- this will be more fully discussed in the next chapter. Consequently, in order to extend the exponential stability results of this section to non-SPR $H(s)$ it proves necessary to restrict $\phi(t)$. We shall next consider a small gain approach.

2.4 RELAXING THE SPR CONDITION

Our aim in this section is to develop stability theorems for (1.1) which allow the relaxation of the SPR condition on $H(s)$ while preserving the exponential stability property. The price of this relaxation will be to restrict the frequency content of the signals $\phi(t)$. As we already know, (1.1) arises as the linearized error equation of our adaptive systems with $H(s)$ being a transfer function arising from a combination of the actual plant being controlled, its nominal parametrized value (about which we linearize), reference model values, etc., while $\phi(t)$ is a regression vector signal composed of plant inputs and outputs. Our reasons for insisting on exponential stability of this system are to ensure robust bounded-input/bounded-output stability properties of the adaptive system when undermodelling, time variations, and disturbances are introduced. Consequently, it is of paramount importance to specify in a robust way the conditions for the exponential stability of the linear homogeneous system itself. Our measure of robustness or "structural stability" of a property is that the property persists under small perturbations. Clearly, satisfaction of persistence of excitation conditions, (3.17) or (3.18) remains valid if $\phi(t)$ is replaced by $\phi(t)+\delta(t)$ where $\delta(t)$ is a signal of arbitrarily small L_p norm. Unfortunately, however, small perturbation of $H(s)$ can invalidate the SPR property since, for example, $H(j\omega)-\delta$ need no longer have nonnegative real part for all ω even though δ is small.

The approach of this section will be to specify more structurally stable conditions for (1.1) to be exponentially stable, with the intention of developing broad robust requirements jointly on signals $\phi(t)$ and transfer functions $H(s)$. The ultimate aim of these results will be to produce guidelines for reference signal selection, adaptation criterion choice, desired closed loop performance, etc., in such a way as to improve robustness of the whole adaptive system. These goals will be pursued in later chapters, but here we shall concentrate on relaxing the SPR condition.

2.4.1 High Frequency Violation of SPR

The deviations from SPR which we shall study here will be where $H(s)$ is an exponentially stable transfer function but the condition $\mathrm{Re}H(j\omega) > 0$ may fail for large ω. We choose this formulation because it admits a straightforward analysis using the small gain theorem and because it reflects both realistic practical effects and the nature of theoretical examples of instability.

Specifically, suppose that $H(s)$ is not SPR but is proper and stable and may be written as

$$H(s) = H_1(s) + sH_2(s) \tag{4.1}$$

where $H_1(s)$ is SPR and, clearly $H(0) = H_1(0)$. Equation (4.1) implies that $H_2(s)$ is stable and strictly proper. We now substitute for $H(s)$ in (1.1) to yield

$$\dot{\theta}(t) = -\epsilon\phi(t)[H_1(s)+sH_2(s)]\{\phi^T(t)\theta(t)\}$$

or, since s is a derivative operator,

$$\begin{aligned}\dot{\theta}(t) = & -\epsilon\phi(t)H_1(s)\{\phi^T(t)\theta(t)\}-\epsilon\phi(t)H_2(s)\{\dot{\phi}^T(t)\theta(t)\} \\ & -\epsilon\phi(t)H_2(s)\{\phi^T(t)\dot{\theta}(t)\}\end{aligned} \tag{4.2}$$

The right hand side above consists of three components. The first corresponds to the right hand side of (1.1) with an SPR transfer function -- Theorem 2.3 describes the exponential stability of this differential equation subject to persistence of excitation of $\phi(t)$. The second and third terms are then best regarded as perturbations to (1.1) whose effects will be analyzed by the small gain theorem. The second term is a parametric perturbation which we shall treat as a feedback interconnection and the third term is a regular perturbation of the complete equation. We proceed by studying first the stability properties of

$$\dot{\theta} = -\epsilon\phi H_1(s)\{\phi^T\theta\} - \epsilon\phi H_2(s)\{\dot{\phi}^T\theta\} \tag{4.3}$$

and then subsequently those of

$$(1+\epsilon\phi H_2(s)\phi^T)\dot{\theta} = -\epsilon M\theta \tag{4.4}$$

where $-\epsilon M\theta$ is the right hand side of (4.3).

2.4.2 Exponential Stability of (4.3)

We write (4.3) as the interconnection of systems

$$\dot{\theta} = -\epsilon\phi H_1(s)\{\phi^T\theta\} + v \tag{4.5}$$

and

$$v = \epsilon\phi H_2(s)\{\dot{\phi}^T w\} \tag{4.6}$$

with unity negative feedback

$$w = -\theta$$

We may then appeal to the small gain theorem in our stability analysis of (4.3).

Theorem 2.3 asserts that, since $H_1(s)$ is SPR, provided $\phi(t)$ satisfies (3.18), the differential equation

$$\dot{\theta} = -\epsilon\phi H_1(s)\{\phi^T\theta\}$$

is exponentially stable so that, denoting the state of $H_1(s)$ by $x(t)$, there exist constants m and λ such that

$$\left\| \begin{bmatrix} x(t) \\ \theta(t) \end{bmatrix} \right\| \leq me^{-\lambda\tau} \left\| \begin{bmatrix} x(0) \\ \theta(0) \end{bmatrix} \right\|, \quad m \geq 1, \ \lambda > 0, \ t \geq 0$$

or, in input-output terms, that

$$\|F(t,\tau)\| \leq me^{-\lambda(t-\tau)}, \quad t \geq \tau \tag{4.7}$$

where $F(.,.)$ denotes the transition matrix of (4.3). Denote $z(t) = [x^T(t) \ \ \theta^T(t)]^T$. From Lemma 2.1(iii) we have that the L_p gain from v to $z(t)$ is bounded above by m/λ and the L_p stability of (4.3) will be assured by the small gain theorem if the gain of (4.6) from w to v is overbounded by λ/m. We now characterize the conditions for this.

Theorem 2.4:

Suppose that $H_1(s)$ is SPR, $\phi(t)$ is persistently exciting, i.e., satisfies (3.18), and further satisfies for all t

$$\|\phi\|_\infty \leq m_1 \tag{4.8}$$

and

$$\|\dot{\phi}\|_\infty \leq m_2 \tag{4.9}$$

Denote the L_p gain of $H_2(s)$ for some particular $p \in [1,\infty)$ by γ_2. Then the solution $\theta(t)$ of (4.3) will be in L_p if

$$\epsilon \frac{m}{\lambda} m_1 m_2 \gamma_2 < 1 \tag{4.10}$$

Proof

From Lemma 2.1 and the theorem conditions the L_p gain of (4.6) is at most $\epsilon m_1 m_2 \gamma_2$. Equation (4.10) is then a restatement of condition (2.8) from the small gain theorem.

□

Having established conditions for L_p stability of (4.3) we next turn to extend this to exponential stability. To do this we use the theory of exponential weighting and write $\theta_a(t) = \theta(t)e^{at}$, $x_a(t) = x(t)e^{at}$, $v_a(t) = v(t)e^{at}$, $z_a(t) = [x_a^T(t)\ \theta_a^T(t)]^T$ for some $0 < a < \lambda$ and, from the bound on $F(t,\tau)$, write

$$|z_a(t)| = \left\| \begin{bmatrix} x_a(t) \\ \theta_a(t) \end{bmatrix} \right\| \leq \int_0^t e^{a(t-\tau)} m e^{-\lambda(t-\tau)} v_a(\tau) d\tau + k(t)$$

where $k(t)$ is due to initial conditions and decays exponentially fast to zero. This yields the L_p gain from v_a to z_a as $m/(\lambda - a)$. The analogous operator from z_a to v_a is

$$v_a(t) = -\epsilon e^{at}\phi(t)H_2(s)\{\dot{\phi}(t)^T e^{-at}\theta_a(t)\}$$

Denote by $\gamma_2(a)$ the shifted L_∞ gain of $H_2(s)$ with impulse response $h_2(t)$

$$\gamma_2(a) = \int_0^{\infty} |h_2(t)| e^{at} dt \tag{4.11}$$

and by $\tilde{\gamma}_2(a)$ the shifted L_2 gain

$$\tilde{\gamma}_2(a) = \max_{\omega} |H_2(j\omega - a)| \tag{4.12}$$

We may now appeal to the small gain theorem once again.

Theorem 2.5:

Suppose that the conditions of Theorem 2.4 hold, constant a is chosen so that $0 < a < \lambda$ and $H_2(s-a)$ has no poles in $\mathrm{Re}(s) \geq 0$. Then $\theta(t)$ satisfies

$$e^{at}\theta(t) \in L_2 \quad \text{if} \quad \epsilon \frac{m}{\lambda - a} m_1 m_2 \tilde{\gamma}_2(a) < 1 \tag{4.13}$$

and

$$e^{at}\theta(t) \in L_\infty \quad \text{if} \quad \epsilon \frac{m}{\lambda - a} m_1 m_2 \gamma_2(a) < 1 \tag{4.14}$$

Further, (4.13) implies that $e^{at}\theta(t) \in L_\infty$ and $e^{at}\theta(t) \to 0$ as $t \to 0$.

Proof

Earlier statements are simple consequences of the small gain theorem while the final part follows from the implication that $e^{at}\theta \in L_2$ and (4.3) forces $e^{at}\dot{\theta}(t) \in L_2$. Thus $\frac{d}{dt}(e^{at}\theta(t)) \in L_2$ and hence the final statement holds.

□

2.4.3 Exponential Stability of (4.4)

Recall that (4.4) is equivalent to (4.2) which, in turn, describes (1.1) with non-S 'R $H(s)$ decomposed as in (4.1). Equation (4.4) is

$$[1+\epsilon\phi(t)H_2(s)\phi^T(t)]\dot{\theta}(t) = -\epsilon M\theta(t)$$

where M is an operator with a prescribed degree of stability:

$$M = \phi(t)H_1(s)\phi^T(t) + \phi(t)H_2(s)\dot{\phi}^T(t)$$

We may identify (4.4) as a regular perturbation of the exponentially stable differential equation (4.3) and, if the new component in the left hand side of (4.4) is small, the exponential stability properties will carry over.

Specifically, we write the operator

$$N = \phi(t)H_2(s)\phi^T(t)$$

and define the new operator

$$\begin{aligned} K &= (I+\epsilon N)^{-1} - I \\ &= -\epsilon N + \epsilon^2 N^2 - \epsilon^3 N^3 + \cdots \end{aligned}$$

provided

$$\epsilon \|N\| < 1 \tag{4.15}$$

We then have a bound on the operator norm

$$\begin{aligned} \|K\| &\leq \epsilon\|N\| + \epsilon^2\|N\|^2 + \epsilon^3\|N\|^3 + \cdots \\ &= \epsilon\|N\|(1-\epsilon\|N\|)^{-1} \end{aligned} \tag{4.16}$$

Notice that condition (4.15) is sufficient for guaranteeing the existence of K and describes formally the conditions for (4.4) to be a simple regular perturbation of (4.3).

We may now write (4.4) as

$$\dot{\theta} = -\epsilon M\theta - \epsilon KM\theta$$

which, in turn, admits description as a feedback system

$$\begin{aligned} \dot{\theta} &= -\epsilon M\theta + v \\ v &= \epsilon KMw \\ w &= -\theta \end{aligned}$$

From Theorem 2.5, we may guarantee that the L^p gain of the first equation is m'/λ', where $\lambda' < \lambda$, and then the small gain theorem may be reapplied to establish stability of (4.4) provided

$$\frac{\epsilon m'}{\lambda'} \|K\| \cdot \|M\| < 1 \tag{4.17}$$

Exponential stability also carries through in a similar manner by considering $\theta_b(t) = e^{bt}\theta(t)$, etc. We shall formally state all these requirements in the next subsection.

2.4.4 Exponential Stability Without the SPR Condition

Having split (1.1) with non-SPR $H(s)$ successively into separate pieces the stability of each of which may be analyzed by small gain methods, it is now time to draw all the conditions together to state the main result concerning (1.1).

Theorem 2.6:

Consider the ordinary differential equation (1.1):

$$\dot{\theta}(t) = -\epsilon\phi(t)H(s)\{\phi^T(t)\theta(t)\}$$

Suppose $H(s)$ is a stable, proper transfer function which may be written as (4.1):

$$H(s) = H_l(s) + sH_2(s)$$

where $H_1(s)$ is SPR and $H_2(s)$ is stable and there is a $b > 0$ such that $H_1(s-b)$ and $H_2(s-b)$ have no poles in $\mathrm{Re}(s) \geq 0$. Further suppose that $\phi(t)$ is persistently exciting and satisfies (4.8) and (4.9)

$$\|\phi\|_\infty \leq m_1 \quad \text{and} \quad \|\dot{\phi}\|_\infty \leq m_2$$

for all t. Then the differential equation will be exponentially stable if all the following conditions hold:

(i) $\epsilon m_1 m_2 \gamma_2(b) \dfrac{m}{\lambda - b} < 1$ (4.18)

(ii) $\epsilon m_1^2 \gamma_2(b) < 1$ (4.19)

(iii) $\epsilon^2 \dfrac{m'}{\lambda' - b} \dfrac{m_1^2\gamma_2(b)}{1-\epsilon m_1^2\gamma_2(b)} [m_1^2\gamma_1(b) + m_1 m_2 \gamma_2(b)] < 1$ (4.20)

where: m and λ are associated with the exponential convergence of the ideal equation and (4.7); m' and λ' are associated with the exponential convergence of (4.3); $\gamma_2(b)$ is the shifted L_∞ gain of $H_2(s)$ defined by (4.11); and, $\gamma_1(b)$ is the equivalent gain for $H_1(s)$. Alternatively, the L_∞ gain $\gamma_i(b)$ may be replaced by $\tilde{\gamma}_i(b)$ defined in (4.12).

Proof

This theorem follows by carrying through the successive conditions of the developments of the previous subsections. Thus (4.18) is (4.10), (4.19) is (4.15) and (4.20) is (4.17) written explicitly in terms of signals and transfer function properties.

□

This theorem describes technical conditions for the attainment of exponential stability of (1.1) with a class of non-SPR $H(s)$. We now turn to consider what the implications are of the inequalities (4.18) - (4.20) and ask, in particular, about their achievability.

The parameters m and λ describe the convergence rate of the differential equation with $H_1(s)$ only appearing, that is, with an SPR transfer function. This equation was the subject of Theorem 2.3 and moreover its convergence rate was the topic of Corollary 2.1. This result states that for given $\phi(t)$ the convergence rate, λ, for small ϵ is proportional to ϵ. Since ϵm_1^2 scales the differential equation (1.1) as ϵ, this result shows that λ is actually proportional to ϵm_1^2 for small ϵ. Further, Corollary 2.1 states that the parameter m for ϵ small is effectively constant in (4.7). Thus the gain of the "ideal" system (4.5) is m/λ which scales as $(m_1^2\epsilon)^{-1}$ for small ϵ and otherwise fixed $\{\phi(t)\}$ which is persistently exciting, i.e., satisfies (3.17) or (3.18) with $Tm_1^2\alpha^{-1}$ fixed. The smallness of ϵ will be more fully explored in the next chapter but here is embodied in the requirement that $\epsilon m_1^2 T < 1$.

Inequality (4.19), for given $\{\phi(t)\}$ and $H_2(s)$, may be satisfied by selecting ϵ sufficiently small. Similarly small ϵ suffices to ensure satisfaction of (4.20) since λ', m' scale like λ, m. The determining inequality therefore arises as (4.18) and the above analysis shows that, replacing $m/(\lambda-b)$ by $k(m_1^2\epsilon)^{-1}$, the operative design parameter once ϵ is chosen small is m_2/m_1. That is, satisfaction of the three inequalities of Theorem 2.6 relies on small ϵ and small m_2/m_1 or, since m_2 and m_1 are the bounds on $|\dot{\phi}(t)|$ and $|\phi(t)|$, low frequency content of the regressor $\phi(t)$. This is entirely consistent with our earlier notions of Subsection 2.3.3 of the SPR condition arising as a sign-preserving operator. If $H(s)$ satisfies (4.1) then it is SPR at low frequences, that is, for input signals having dominant low frequency content the signs of input and output are the same in the integral sense of (3.10). Restricting ϵ to be small causes $\theta(t)$ to be a low frequency signal so that $\phi^T(t)\theta(t)$ is dominantly low frequency and the SPR condition need only be valid in this range.

Another preliminary question to ask is: How does one choose $H_1(s)$ in (4.1)? It is clear that $H_1(s)$ is nonunique since its only specification is that $H_1(0) = H(0)$ and that $H_1(s)$ is SPR. Further, the particular choice of H_1 affects the constants $\gamma_1(b)$, $\gamma_2(b)$, m, λ, m', λ' in the Theorem above. We really do not attempt to answer this question fully here but do point out that our experience with examples has shown that some choices of H_1 lead to very much less conservative stability requirements. The aim in choosing H_1 is clearly to produce the least restrictive subsequent assumptions but it is difficult to formulate rules for this and so we shall, for the moment at least, regard H_1 and H_2 as fixed and examine the effects and means of varying the remaining parameters ϵ, m, λ, m_1, and m_2. We should remark here, however, that the formulation using (4.1) does force $H_2(s)$ to be stable and strictly proper and hence always to have finite L_2 gain -- in a sense high frequency deviations of $H(s)$ from SPR $H_1(s)$ are embodied in the factor s multiplying $H_2(s)$. Since H is proper $H_2(s)$ must have at least a first-order asymptotic roll-off at high frequencies.

In the preceding developments of exponential stability the implicit presumption of persistence of excitation of $\phi(t)$ was made but, evidently from (3.17) or (3.18), one cannot arbitrarily decrease the frequency content of $\phi(t)$ without taking pains to ensure the continued satisfaction of this persistence condition. The problem here is that as the frequency content of $\phi(t)$ is decreased the ability of ϕ to be persistently spanning at level α over all intervals of length T is impaired since $\phi(t)\phi^T(t)$ is a rank one matrix and restricting the frequency content limits the variation of this matrix over fixed intervals. One way to achieve satisfaction of the active inequality (4.18) by reducing m_2/m_1 is to replace $\phi(t)$ by $\phi(t/l)$ for $l > 1$, i.e., to stretch the time scale. This has the effect of replacing m_1, m_2, α, and T by m_1, m_2/l, αl, lT with the result that, provided ϵlT is still small, the degree of stability is unchanged but the left hand side of (4.18) has been reduced by a factor of l^{-1}. This is illustrated well by considering, for example, the signal $\phi(t) = [\cos\omega t, \sin\omega t]^T$ in (3.17) and letting ω decrease while keeping $\frac{\epsilon}{\omega}$ small.

Time scale stretching may not always be easily achieved when $\phi(t)$ is derived as an ARMA regression vector, since the plant frequency response comes into play. Conditions for ensuring persistence of excitation of ϕ by manipulating plant inputs alone will be the subject of the next section.

2.5 PERSISTENCE OF EXCITATION

We have presented persistence of excitation conditions at several stages in the analysis so far, such as (3.17) and (3.18), and these have referred specifically to the regression vector $\phi(t)$ and its spanning properties uniformly over time. A brief inspection of (1.1) or of its simplified version (3.6) shows that nondegeneracy of $\phi(t)$ is a condition which constrains the linearized adaptation algorithm only to have a single asymptotically stable equilibrium point at $\theta = 0$. Uniformity of this nondegeneracy over time (or persistence of excitation) implies uniform convergence. In terms of parameter estimation theory, persistence of excitation is a uniform identifiability condition allowing identification of the correct parameter value from measured regressors and errors.

As will be reiterated later in Chapters 4 and 5, one frequently deals with ARMA models of linear systems in continuous-time

$$y^{(n)}(t)+a_1y^{(n-1)}(t)+\cdots+a_ny(t) = b_1u^{(n-1)}(t)+\cdots+b_nu(t) \tag{5.1}$$

or in discrete-time

$$y_k + c_1y_{k-1} + \cdots + c_ny_{k-n} = d_1u_{k-1} + \cdots + d_nu_{k-n} \tag{5.2}$$

which may be written in the polynomial forms

$$y(s) = A^{-1}(s)B(s)u(s) \tag{5.3}$$

or

$$y(z) = C^{-1}(z)D(z)u(z)$$

respectively, where s is the Laplace transform variable and z is the $Z-$transform variable. We shall concentrate here on the continuous-time problem and address discrete-time within Chapter 5. Depending on the parametrization used for (5.1) the regression vector could (in theory) be chosen as a linear combination of u and its $(n-1)$ derivatives and y and its $(n-1)$ derivatives. More realistically, the regression vector is chosen to be composed of proper stable filterings of u and y, which could be thought of as approximating these derivatives (albeit crudely). Specifically, denote by $u(s)$ and $y(s)$ the Laplace transforms of $u(t)$ and $y(t)$ and denote by $\overline{u}^{(l)}(t)$ and $\overline{y}^{(l)}(t)$ the signals whose Laplace transforms are

$$\overline{u}^{(l)}(s) = h^{(l)}(s)u(s) \tag{5.4}$$

$$\overline{y}^{(l)}(s) = h^{(l)}(s)y(s)$$

for $0 \leq i \leq n-1$, where $\{h^{(i)}(s)\}$ is a collection of proper stable transfer functions with precise form given below, and $\bar{u}^{(i)}(0)$ and $\bar{y}^{(i)}(0)$ are in some finite ball. Then one may take

$$\phi^T(t) = (\bar{y}^{(n-1)}(t), \bar{y}^{(n-2)}(t), \ldots, \bar{y}^{(0)}(t), \bar{u}^{(n-1)}(t), \ldots, \bar{u}^{(0)}(t)) \tag{5.5}$$

since -- provided the $h^{(i)}(s)$ are suitably selected -- $(\bar{y}^{(i)}(t), \bar{u}^{(i)}(t))$ satisfy an ARMA system of equations identical to (5.1) modulo initial condition transients which decay to zero like the degree of stability of $h^{(i)}(s)$. The stability and properness of these $h^{(i)}(s)$ are important to ensure, respectively, the decay of transients, and the realizability of $\phi(t)$ from measurements of the true signals $u(t)$ and $y(t)$ which need not be completely smooth nor arbitrarily differentiable -- the exact class of allowable $u(t)$ will be discussed briefly later. Some typical choices for $h^{(i)}(s)$ in (5.4) are $(s+\alpha)^{-i}$ or $s^i(s+\alpha)^{-m}$ for $\alpha > 0$ and some m.

We ask the question: What are sufficient conditions on $u(t)$ to ensure persistence of excitation of $\phi(t)$ given by (5.5)? This is an important question in adaptive systems since, as we shall see shortly, $\phi(t)$ is not directly able to be manipulated to achieve persistence of excitation but may be indirectly controlled by modifying other signals such as reference trajectories and test signals which may be chosen by the designer of the adaptive system.

We derive firstly a preliminary result:

Lemma 2.7:

Let $v(t)$ be an $n-1$ times differentiable $m-$vector function and let $x(t)$ be a $n-$vector function defined by

$$\dot{x}(t) = Ax(t) + Bv(t) \ , \ x(0) = x_0 \tag{5.6}$$

$[A,B]$ being a controllable pair. Denote

$$V(t) = (v^T(t)\dot{v}^T(t) \ldots v^{(n-1)T}(t))^T$$

If there exist strictly positive constants S and γ such that, for all

$t \geq 0$,

$$\frac{1}{S}\int_t^{t+S} V(\tau)V^T(\tau)d\tau \;>\; \gamma I \;>\; 0 \tag{5.7}$$

then there exists a strictly positive constant β such that, for all $t \geq 0$,

$$\frac{1}{S}\int_t^{t+S} x(\tau)x^T(\tau)d\tau \;>\; \beta I \;>\; 0 \tag{5.8}$$

Proof

Choose a particular value of t. Then for any scalar function $y_\alpha(\tau)$, $t \leq \tau \leq t+S$, we have from the Cauchy-Schwarz inequality

$$\int_t^{t+S} x(\tau)x^T(\tau)d\tau \int_t^{t+S} y_\alpha^2(\tau)d\tau \;\geq\; \int_t^{t+S} x(\tau)y_\alpha(\tau)d\tau \int_t^{t+S} x^T(\tau)y_\alpha(\tau)d\tau$$

We shall choose functions $y_\alpha(t)$ which have desirable properties. In particular

$$y_\alpha(\tau) \;=\; \sum_{l=0}^{n} (-1)^l p_l \Phi_\alpha^{(l)}(\tau)$$

where: $p(\lambda) = p_0 + p_1\lambda^2 + \cdots + p_n\lambda^n$, $p_n = 1$ is the characteristic (or any annihilating) polynomial of the matrix A, i.e., $p(A) = 0$; $\Phi_\alpha(\tau)$ is any n times differentiable approximation to the Dirac delta function at $\tau = t+\alpha$ which satisfies $\Phi_\alpha^{(l)}(t-\epsilon) = \Phi_\alpha^{(l)}(t+S+\epsilon) = 0$ for $i \leq n$ where $\Phi_\alpha^{(l)}$ is the ith derivative of Φ_α and ϵ is an arbitrarily small number. Clearly

$$C \;=\; \int_t^{t+S} y_\alpha^2(\tau)d\tau \;>\; 0$$

Consider now

$$\int_{t-\epsilon}^{t+S+\epsilon} x(\tau)y_\alpha(\tau)d\tau \;=\; \sum_{l=0}^{n}(-1)^l p_l \int_{t-\epsilon}^{t+S+\epsilon} x(\tau)\Phi_\alpha^l(\tau)d\tau$$

$$=\; \sum_{l=0}^{n} p_l \int_{t-\epsilon}^{t+S+\epsilon} x^{(l)}(\tau)\Phi_\alpha(\tau)d\tau$$

after integration by parts and using the boundary conditions on Φ_α. Next we use (5.6) to evaluate

$$\sum_{i=0}^{n} p_i x^{(i)}(\tau) = p_0 x + p_1(Ax+Bv) + p_2(A^2x+ABv+B\dot{v})$$

$$+ \cdots + p_n(A^n x+A^{n-1}Bv+ \cdots +Bv^{(n-1)})$$

and recall that $p(\cdot)$ is the characteristic polynomial of A to write

$$\sum_{i=0}^{n} p_i x^{(i)}(\tau) = [(\sum_{i=1}^{n} p_i A^{i-1}B) \cdots (\sum_{i=j}^{n} p_i A^{i-j}B) \cdots (p_{n-1}B+AB)\; B]V(\tau)$$

$$=: QV(\tau)$$

Thus

$$\int_{t-\epsilon}^{t+S+\epsilon} x(\tau)y_\alpha(\tau)d\tau = \int_{t-\epsilon}^{t+S+\epsilon} QV(\tau)\Phi_\alpha(\tau)d\tau = QV(t+\alpha) + e$$

since Φ_α is an arbitrary approximation to a delta function. The term e is the arbitrarily small error term. Therefore (modulo the arbitrarily small correction term)

$$\int_{t-\epsilon}^{t+S+\epsilon} x(\tau)x^T(\tau)d\tau \geq \frac{1}{C}\, QV(t+\alpha)V^T(t+\alpha)Q^T$$

and integrating both sides with respect to α from 0 to S we have

$$\int_{t-\epsilon}^{t+S+\epsilon} x(\tau)x^T(\tau)d\tau \geq \frac{1}{S}\,\frac{1}{C}\, Q \int_{t-\epsilon}^{t+S+\epsilon} V(\tau)V^T(\tau)d\tau Q^T$$

The matrix Q above has full row rank if and only if $[A,B]$ is controllable and the result follows by continuity of x.

□

These results are easily extended to outputs of linear systems subject to reachability requirements.

Lemma 2.8:

Subject to the hypothesis of Lemma 2.7 let $y(t)$ be defined by

$$y(t) = Cx(t)$$

Then $v(t)$ satisfying (5.7) implies that there exists a strictly positive constant α such that for all $t \geq 0$

$$\frac{1}{S}\int_t^{t+S} y(\tau)y^T(\tau)d\tau > \alpha I \tag{5.9}$$

if the triple $[A,B,C]$ is output reachable, i.e., if the matrix

$$[CB \quad CAB \quad \cdots \quad CA^{n-1}B]$$

has full row rank. Further, if $y(t)$ is generated as

$$y(t) = Cx(t) + Dv(t)$$

where the matrix

$$[D \quad CB \quad CAB \quad \cdots \quad CA^{n-1}B]$$

has full row rank and $v(t)$ is n times differentiable, then (5.9) is implied by

$$\frac{1}{S}\int_t^{t+S} \bar{V}(\tau)\bar{V}^T(\tau)d\tau > \lambda I > 0 \tag{5.10}$$

for some positive S and λ and all t, where

$$\bar{V}(t) = (V^T(t)\ v^{(n)T}(t))^T$$

Proof

Multiplying the last inequality of the proof of Lemma 2.7 by C on the left and C^T on the right, and factoring the matrix Q as

$$Q = [B \ AB \ A^2B \ \cdots \ A^{n-1}B] \begin{bmatrix} p_1I & p_2I & \cdots & p_{n-1}I & p_nI \\ p_2I & p_3I & \cdots & p_nI & 0 \\ \vdots & \vdots & & & \\ p_{n-1}I & p_nI & & & \\ p_nI & 0 & & & 0 \end{bmatrix}$$

where $p_n = 1$ by assumption, the first result follows directly. To establish the remaining part of the result we return to the proof of Lemma 2.7 and evaluate quantities like

$$\frac{1}{S} \int_{t-\epsilon}^{t+S+\epsilon} y(\tau) y_\alpha(\tau) d\tau = \bar{Q}\bar{V}(t+\alpha) + e$$

where

$$\bar{Q} = [D \ CB \ CAB \ \cdots \ CA^{n-1}B] \begin{bmatrix} p_0I & p_1I & \cdots & p_{n-2}I & p_{n-1}I & p_nI \\ p_1I & p_2I & \cdots & p_{n-1}I & p_nI & 0 \\ p_2I & p_3I & \cdots & p_nI & 0 & 0 \\ \vdots & \vdots & & & & \\ p_{n-1}I & p_nI & & & & \\ p_nI & 0 & & & 0 & 0 \end{bmatrix}$$

□

We are now in a position to approach more directly the full question of achieving persistence of excitation of regression vectors from input signals only. If our plant model is an ARMA process as in (5.1) our regressor $\phi(t)$ is given by (5.5) and we may reform our requirements with

Theorem 2.7:

Suppose that $y(t)$ and $u(t)$ are related by the ARMA model (5.1) with the polynomials $A(s) = s^n + a_1 s^{n-1} + \cdots + a_n$ and $B(s) = b_1 s^{n-1} + \cdots + b_n$ being relatively prime. If there exist positive constants α and S such that

$$U(t) = (\bar{u}(t) \ \bar{u}^{(1)}(t) \cdots \bar{u}^{(2n-1)}(t))^T$$

where $\bar{u}^{(l)}(t)$ is given by (5.4) and $h^{(l)}(s) = s^l(s+\gamma)^{-2n}$, $\gamma > 0$ satisfies

$$\frac{1}{S}\int_t^{t+S} U(\tau)U^T(\tau)d\tau > \alpha I \tag{5.11}$$

for all $t \geq 0$, then $\phi(t)$ given by (5.5) satisfies

$$\frac{1}{S}\int_t^{t+S} \phi(\tau)\phi^T(\tau)d\tau > \beta I \tag{5.12}$$

for some positive constant β and all $t \geq t_0$ for some finite t_0.

Proof

Modulo transient terms which decay like $e^{-\gamma t}$, for each i, $\bar{u}^{(l)}(t)$ and $\bar{y}^{(l)}(t)$ are related by the same ARMA model as $u(t)$ and $y(t)$, (5.1). Write a minimal state space realization of

$$A^{-1}(s)B(s) = H(sI-F)^{-1}G$$

where, since $A(s)$ and $B(s)$ are coprime, F is $n \times n$ and we may take (F,G,H) in observable canonical form.

$$F = \begin{bmatrix} -a_1 & 1 & 0 & \cdots & 0 \\ -a_2 & 0 & 1 & \cdots & 0 \\ \vdots & & & & \\ -a_n & 0 & 0 & \cdots & 0 \end{bmatrix}, \quad G = \begin{bmatrix} b_1 \\ b_2 \\ \vdots \\ b_n \end{bmatrix}$$

$$H = [1 \ 0 \ 0 \ \cdots \ 0]$$

This yields immediately a minimal state-space realization of the map from $W(t) = (\bar{u}(t)\, \bar{u}^{(1)}(t) \cdots \bar{u}^{(n-1)}(t))^T$ to $\phi(t)$ as $(\bar{F},\bar{G},\bar{H},\bar{J})$ with

$$\bar{F} = F \oplus F \oplus \cdots \oplus F \;, \quad \bar{G} = G \oplus G \oplus \cdots \oplus G$$

$$\bar{H} = \begin{bmatrix} H \oplus H \oplus \cdots \oplus H \\ 0 \end{bmatrix} , \quad \bar{J} = \begin{bmatrix} 0 \\ I \end{bmatrix}$$

where $\oplus$ denotes Kronecker (or direct) sum.

The output reachability of this system is immediate from the definition of H and the controllability of $[F,G]$. Further, from the structure of $\bar{F}$, it is clear that the characteristic polynomial $p(\cdot)$ of F is an annihilating polynomial of $\bar{F}$ and the degree of p is n. Following through the proofs of Lemmas 2.7 and 2.8, $\phi(t)$ will satisfy (5.12) provided

$$\bar{Q} \int_t^{t+S} \bar{W}(\tau)\bar{W}^T(\tau)d\tau \bar{Q}^T > \delta I$$

for some $\delta > 0$ and all t where $\bar{Q}$ is the analogue of Q appearing in the proof of Lemma 2.8 constructed with $(\bar{F},\bar{G},\bar{H},\bar{J})$ and

$$\bar{W}(\tau) = (W^T(\tau) \cdots W^{(n)T}(\tau))^T$$

Hence, noting that $\bar{u}^{(i+1)}(\tau) = \dot{\bar{u}}^{(i)}(\tau)$, we have

$$\bar{W}(\tau) = \bar{M}U(\tau)$$

with

$$\bar{M} = \begin{bmatrix} I_n & & O_{n,n} \\ & \vdots & \\ [O_{n,i} & I_n & O_{n,n-i}] \\ & \vdots & \\ O_{n,n} & & I_n \end{bmatrix} \quad i = 0,\ldots,n$$

where $O_{n,i}$ denotes a zero matrix of n rows and i columns.

Hence, $\phi(t)$ satisfies (5.12) if

$$\overline{Q}\overline{M}\int_{t}^{t+S} U(\tau)U^T(\tau)d\tau\overline{M}^T\overline{Q}^T > \delta I$$

which, in turn, follows from (5.11) provided that the matrix $\overline{Q}\overline{M}$ has full row rank. Now

$$\overline{Q} = [\overline{J}\ \overline{HG}\ \overline{HFG}\cdots\overline{HF^{n-1}G}]\begin{bmatrix} p_0 I & p_1 I & \cdots & p_{n-1}I & p_n I \\ p_1 I & p_2 I & \cdots & p_n I & 0 \\ \vdots & \vdots & & & \\ p_n I & 0 & 0 & 0 & 0 \end{bmatrix}$$

$$= \begin{bmatrix} \overline{Q}_0 & \overline{Q}_1 & \cdots & \overline{Q}_{n-1} & 0 \\ p_0 I & p_1 I & \cdots & p_{n-1}I & p_n I \end{bmatrix}$$

from the gross structure of $\overline{F},\overline{G},\overline{H},\overline{J}$ where

$$\overline{Q}_i = (\sum_{j=i+1}^{n} p_j HF^{j-i-1}G) \oplus \cdots \oplus (\sum_{j=i+1}^{n} p_j HF^{j-i-1}G)$$

$$= Q_i \oplus \cdots \oplus Q_i$$

Hence

$$\overline{Q}\overline{M} = \begin{bmatrix} Q_0 & Q_1 & Q_2 & \cdots & Q_{n-1} & 0 & \cdots & 0 & 0 \\ 0 & Q_0 & Q_1 & \cdots & Q_{n-2} & Q_{n-1} & \cdots & 0 & 0 \\ \vdots & & & & & & & & \\ 0 & 0 & 0 & \cdots & 0 & Q_0 & \cdots & Q_{n-1} & 0 \\ p_0 & p_1 & p_2 & \cdots & p_{n-1} & p_n & \cdots & 0 & 0 \\ 0 & p_0 & p_1 & \cdots & p_{n-2} & p_{n-1} & \cdots & 0 & 0 \\ \vdots & & & & & & & & \\ 0 & 0 & 0 & \cdots & 0 & p_0 & \cdots & p_{n-1} & p_n \end{bmatrix}$$

i.e., $\overline{Q}\overline{M}$ is the resultant matrix of the polynomials $q(s)$ and $p(s)$ whose coefficients are Q_i and p_i, respectively. Since $[F,G,H]$ is a minimal state variable realization of $A^{-1}(s)B(s)$, which is strictly proper, we have

$$B(s) = A(s)\{HGs^{-1} + HFGs^{-2} + HF^2Gs^{-3} + \cdots\}$$

and $A(s) = \det(sI - F)$. From this equality it follows that $A(s) = p(s)$ and $B(s) = q(s)$ which by assumption are coprime. Hence $\overline{Q}\overline{M}$ has full rank and the theorem is proved.

□

This theorem presents conditions on the input signal to a plant to be identified which result in the persistence of excitation of the complete regression vector composed of inputs and outputs. The condition derived, that $U(t)$ satisfy (5.11) where U contains $\bar{u}$ and its first $2n-1$ derivatives, is a richness constraint which has a useful interpretation where u is given as a sum of distinct sinusoids. In this case U can satisfy (5.11) if and only if u contains at least n distinct frequencies ($2n$ complex frequencies). Thus, the spectral content of u needs to be adequate. If fewer frequencies are included in u then it is possible to find constant vectors which are orthogonal to $\phi(t)$ or $U(t)$. Thus, there are directions in $\mathbb{R}^n$ in which the components of $\theta(t)$ are not uniformly decrescent. There are other interpretations of this "input richness" requirement concerned with identifiability of transfer functions, etc., but we do not pursue these here.

There are several areas for extension of these results on persistence conditions to obtain more general or more specialized descriptions. We first note that frequently an upper bound constraint in (5.12) is desired and that this can be achieved by an upper bound on (5.11) and an assumption of stability of $A(s)$. Second, the appearance of input derivatives in Lemma 2.6 is somewhat alarming since the filters we consider later are all proper and assumptions of differentiability of the input should not be needed. We have resolved this point by considering filtered signal $\bar{u}^{(l)}(t)$ but more direct (and less transparent) approaches may be broached, typically by restricting $u(t)$ to belong to a class of piecewise differentiable functions. Equally, different choices for filters allow us to use regressors $\phi(t)$ which are not filtered to as high a degree as those of Theorem 2.7.

Finally, we have already mentioned that consideration of almost periodic or sums-of-sinusoids inputs helps to clarify thinking and this approach can be carried through to encompass inputs possessing power spectral densities or related classes of functions.

The results of Theorem 2.7 were derived to allow us to devolve persistence requirements from plant inputs and outputs to input signals alone. In some adaptive control problems we do not necessarily have the ability to ascribe directly the input properties but may only manipulate more indirect quantities like reference signals or desired output trajectories. (This contrasts to many applications of adaptive identification, however, where system inputs are at the disposal of the operator.) It is possible to establish the requisite persistenc of excitation relationships for such adaptive control problems.

The notion of requiring sufficiently rich reference signals in adaptive control problems highlights a dichotomy of adaptive regulation. The control desire to maintain output signals at constant reference levels is in contravention of the adaptation requirement of persistence of excitation which reduces the achieved control performance. There is a need to formulate a compromise between these competing objectives and this leads to ad hoc implementation changes such as adapting only when recent signal levels satisfy persistence conditions and to formulation of "dual" adaptive control methods via explicit optimization techniques. In any event, it appears to a certain extent, as if persistence of excitation is the price of robustness in adaptive control.

2.6 DISCRETE-TIME RESULTS

The bulk of our analysis so far has concentrated on the continuous-time error system equations (1.1) or (3.2) rather than the discrete-time version (1.2). Here we pursue the presentation of discrete-time equivalents of our earlier results. We shall in general not prove these here as, typically, the proofs are almost identical to or easy modifications of the continuous versions. There are some new issues, however, such as normalization and strict causality, which do arise and these will be addressed at the appropriate time.

2.6.1 Discrete-Time Algorithm Descriptions, Normalization, Causality

Recall that the object of our study is (1.2)

$$\theta_{k+1} = \theta_k - \epsilon\phi_k H(z)\{\phi_k^T\theta_{k+1}\}$$

To help focus our ideas, consider the simple scalar difference equation

$$x_{k+1} = a_k x_k \tag{6.1}$$

If $\{a_k\}$ is a sequence of scalars with magnitude uniformly less than one then (6.1) will be exponentially stable, since the solution is clearly

$$x_k = a_{k-1}a_{k-2} \cdots a_1 a_o x_0$$

Notice that the sign of $\{a_k\}$ is immaterial for stability but the magnitude is of paramount importance. The discrete-time variant of the simple example (3.6) is the normalized least mean square (LMS) algorithm

$$\theta_{k+1} = \theta_k - \epsilon \frac{\phi_k}{1+\epsilon\phi_k^T\phi_k} \phi_k^T\theta_k \tag{6.2}$$

which, like (3.6), is a rank-one updating of the vector θ.

We may write (6.2) as a homogeneous linear equation

$$\theta_{k+1} = \left(I - \frac{\epsilon}{1+\epsilon\phi_k^T\phi_k} \phi_k\phi_k^T \right)\theta_k \tag{6.3}$$

which admits direct comparison to (6.1). Alternatively, by

premultiplying (6.2) by ϕ_k^T we identify

$$\phi_k^T\theta_k = (1 + \epsilon\phi_k^T\phi_k)\phi_k^T\theta_{k+1}$$

so that (6.2) is equivalent to the non-strictly causal algorithm

$$\theta_{k+1} = \theta_k - \epsilon\phi_k(\phi_k^T\theta_{k+1}) \tag{6.4}$$

At this point some comments should be made. (i) The stability requirement for (6.1) indicates that magnitude properties of the matrix in the update are important. In particular, we know from linear time-invariant systems of the connection between stability and matrix eigenvalues having magnitudes less than one. Examining (6.3) we see that the division by $(1+\epsilon\phi_k^T\phi_k)$ in the algorithm assures that the norm of θ_k is nonincreasing whatever the value of ϕ_k. (ii) Some adaptive algorithms, such as the LMS scheme from adaptive filtering, do not include this normalization and for their application require ϕ_k to be otherwise guaranteed bounded. Hence, these unnormalized algorithms are not suitable for adaptive control and find application more readily in signal processing problems. (iii) We note that, while (6.3) appears outwardly dissimilar to the continuous-time algorithm (3.6), the equivalent form (6.4) recovers the form of the continuous version at the expense of being non-strictly causal. The distinction between the two forms of the parameter update is connected with the computation of error signals prior to parameter update -- *a priori* errors -- and computation subsequent to parameter update -- *a posteriori* errors. More will be said about this in Chapter 5.

Equation (1.6) is the more general analogue of (6.4). Its strictly causal representation -- analogous to (6.2) -- requires the introduction of the inverse of an operator but also exhibits this normalization:

$$\theta_{k+1} = \theta_k - \epsilon\phi_k[H(z)\{1+\epsilon\phi_k^T\phi_k H(z)\}^{-1}]\phi_k^T\theta_k$$

In state-variable form, however, the description is somewhat more complicated but does admit the following causally implementable form. Denote by (A,b,c,d) a minimal state variable realization of $H(z)$, i.e.,

$$H(z) = d + c(zI-A)^{-1}b$$

Let us introduce variables η_k, v_k to denote $\phi_k^T\theta_{k+1}$, $H(z)\{\phi_k^T\theta_{k+1}\}$. Then an equivalent causal description of (1.2) is given by

$$x_{k+1} = Ax_k + b\eta_k \tag{6.5}$$

$$v_k = c^Tx_k + d\eta_k \tag{6.6}$$

$$\eta_k = \phi_k^T\theta_k - \epsilon\phi_k^T\phi_k v_k \tag{6.7}$$

$$\theta_{k+1} = \theta_k - \epsilon\phi_k v_k \tag{6.8}$$

In turn, we may reduce this by substituting (6.7) into (6.5) and (6.6) to yield

$$v_k = \frac{1}{1+d\epsilon\phi_k^T\phi_k} \cdot c^Tx_k + \frac{d}{1+d\epsilon\phi_k^T\phi_k}\,\phi_k^T\theta_k$$

which may be substituted into the other resulting equation for x_{k+1} and into (6.8) to give the following strictly causal form:

$$\begin{bmatrix} x_{k+1} \\ \theta_{k+1} \end{bmatrix} = \begin{bmatrix} A - \dfrac{\epsilon\phi_k^T\phi_k}{1+d\epsilon\phi_k^T\phi_k}\,bc^T & \dfrac{1}{1+d\epsilon\phi_k^T\phi_k}\,b\phi_k^T \\[2ex] \dfrac{-\epsilon}{1+d\epsilon\phi_k^T\phi_k}\,\phi_k c^T & I - \dfrac{\epsilon d}{1+d\epsilon\phi_k^T\phi_k}\,\phi_k\phi_k^T \end{bmatrix} \begin{bmatrix} x_k \\ \theta_k \end{bmatrix} \tag{6.9}$$

Equation (6.9) may be viewed as a state-variable description of (1.2). Note the similarity to the corresponding continuous-time equation (3.2) and the appearance of normalizing terms above.

2.6.2 Exponential Stability and the SPR Condition

We are now in a position to begin the analysis of (1.2) via its state-variable description (6.9) and so state the discrete-time equivalent of the Lyapunov stability result Lemma 2.2.

Lemma 2.9:

Consider the difference equation

$$x_{k+1} = F_k x_k \tag{6.10}$$

with transition matrix $\Phi(k+n,k) = F_{k+n-1}F_{k+n-2} \cdots F_{k+1}F_k$. Suppose there exists a positive definite symmetric constant matrix $\bar{P}$ such that

$$F_k^T \bar{P} F_k - \bar{P} = -N_k N_k^T \tag{6.11}$$

for some matrix sequence $\{N_k\}$ and all k. Then (6.10) is stable in the sense of Lyapunov.

If further, the pair $[F_k, N_k]$ is uniformly completely observable, i.e. there exist constants $T > 0$, $\beta > \alpha > 0$ such that for all k

$$\infty > \beta I \geq \sum_{i=0}^{T-1} \Phi^T(k+i,k) N_{k+i} N_{k+i}^T \Phi(k+i,k) \geq \alpha I > 0 \tag{6.12}$$

then (6.10) is exponentially stable.

Finally, the condition of uniform complete observability of $[F_k, N_k]$ may be replaced by the same condition on $[F_k - K_k N_k^T, N_k]$ for bounded $\{K_k\}$ and the same conclusion holds.

□

The proof of this result follows almost identically to the continuous-time Lemmas 2.2 and 2.3.

As an example of the application of this lemma it is useful to consider the stability of (6.3).

Theorem 2.8:

The difference equation (6.3)

$$\theta_{k+1} = \left(I - \frac{\epsilon}{1+\epsilon\phi_k^T\phi_k} \phi_k\phi_k^T \right)\theta_k$$

is globally exponentially stable for any $\epsilon > 0$ provided there exist positive constants T, α, β such that for all k

$$\infty > \beta I \geq \sum_{i=0}^{T-1} \phi_k\phi_k^T \geq \alpha I > 0$$

Proof

Take $\bar{P} = I$ to yield

$$N_k = \frac{[\epsilon(2+\epsilon\phi_k^T\phi_k)]^{1/2}}{1+\epsilon\phi_k^T\phi_k} \cdot \phi_k$$

and choosing $K_k = \epsilon^{1/2}[2+\epsilon\phi_k^T\phi_k]^{-1/2}\phi_k$ gives $F_k - K_kN_k^T = I$ and the result then follows directly since the observability gramian (6.12) of $[F_k - K_kN_k^T, N_k]$ is constrained to lie between bounded multiples of the sum in the theorem statement.

□

From this result we see from the condition that ϵ be positive that once again the sign preservation of $\phi_k^T\theta_k$ is crucial for stability and so it is not surprising that discrete SPR functions should arise here as well.

Definition: A discrete transfer function $H(z)$ is positive real (PR) if

(i) $H(z)$ is analytic in $|z| > 1$ and the poles of $H(z)$ on $|z| = 1$ are simple, with nonnegative residue; and

(ii) $H(e^{j\omega}) + H(e^{-j\omega}) \geq 0$ for all real ω at which $H(e^{j\omega})$ exists.

The discrete transfer function is strictly positive real if $H(\rho z)$ is positive real for some positive $\rho < 1$.

The following sign preservation property, equivalent to (3.10), holds for discrete PR operators, i.e., there exists a constant γ depending on initial conditions in H such that for any function $x \in l_{2e}$ and all t:

$$\sum_{k=0}^{t} x_k (Hx)_k \geq -\gamma$$

Lemma 2.10: The Discrete Positive Real Lemma

Let $H(z)$ be a real rational function of the complex variable z, with $|H(\infty)| < \infty$. Let $\{A,b,c,d\}$ be a minimal realization of $H(z)$, i.e.,

$$H(z) = d + c^T(zI-A)^{-1}b$$

Then $H(z)$ is SPR if and only if there exist real matrices P and L with P positive definite and constants w and γ such that

$$\begin{aligned} A^TPA - P &= -LL^T - \gamma^2 I \\ A^TPb &= c + wL \\ w^2 &= 2d - b^TPb \end{aligned} \tag{6.13}$$

□

We may appeal to this discrete version of the Popov-Kalman-Yakubovich Lemma (Lemma 2.5) analogously to the continuous case to derive new state-variable realizations as is done in Lemma 2.6. Alternatively, we may directly apply the discrete Lyapunov methods of Lemma 2.8 to the state equation (6.9), as follows. We note that the proof here is conceptually very similar to that of Theorem 2.7 but will be presented in full because of the result's importance.

Theorem 2.9:

Consider the ordinary difference equation (1.2)

$$\theta_{k+1} = \theta_k - \epsilon\phi_k H(z)\{\phi_k^T\theta_{k+1}\}$$

where $H(z)$ is a discrete strictly positive real rational transfer function. Then this equation is stable. Further, it is exponentially stable if there exist positive constants T, α, and β such that for all k

$$\infty > \beta I \geq \sum_{l=0}^{T-1} \phi_{k+l}\phi_{k+l}^T \geq \alpha I > 0 \tag{6.14}$$

Additionally, for small ϵ the rate of convergence of θ_k to zero (or degree of stability of (1.2)) is like $(1-m\epsilon)^k$ for some constant m.

Proof

The error system (6.5)-(6.8) may be depicted as a feedback interconnection of (passive) blocks as in Figure 2.4. Since $H(z)$ is SPR we see from (6.13) that $d > 0$ and so we may choose a positive constant δ so that the same error system may be drawn as in Figure 2.5 with the upper loop still SPR as may be seen by considering the discrete analogue of (3.19) above. (Note that the error system is unaltered but this technical device is necessary for later analysis.) We now rewrite (6.5) - (6.8) with new variables, viz.,

$$\bar{\eta}_k = \eta_k - \delta v_k$$

The equations become

$$x_{k+1} = \bar{A}x_k + \bar{b}\bar{\eta}_k$$

$$v_k = \bar{c}^T x_k + \bar{d}\bar{\eta}_k$$

$$\bar{\eta}_k = \phi_k^T\theta_k - (\delta+\epsilon\phi_k^T\phi_k)v_k$$

$$\theta_{k+1} = \theta_k - \epsilon\phi_k v_k$$

where $\bar{A} = (A+\delta(1-d\delta)^{-1}bc^T)$, $\bar{b} = (1-d\delta)^{-1}b$, $\bar{c} = (1-d\delta)^{-1}c$, $\bar{d} = (1-d\delta)^{-1}d$. Equation (6.9) now may be written

$$\begin{bmatrix} x_{k+1} \\ \theta_{k+1} \end{bmatrix} = \begin{bmatrix} \bar{A} - \dfrac{\delta+\epsilon\phi_k^T\phi_k}{1+\bar{d}\epsilon\phi_k^T\phi_k+\bar{\delta}\delta}\,\bar{b}\bar{c}^T & \dfrac{1}{1+\bar{d}\epsilon\phi_k^T\phi_k+\bar{d}\delta}\,\bar{b}\phi_k^T \\ \dfrac{-\epsilon}{1+\bar{d}\epsilon\phi_k^T\phi_k+\bar{d}\delta}\,\phi_k\bar{c}^T & I - \dfrac{\epsilon\bar{d}}{1+\bar{d}\epsilon\phi_k^T\phi_k+\bar{d}\delta}\,\phi_k\phi_k^T \end{bmatrix} \begin{bmatrix} x_k \\ \theta_k \end{bmatrix}$$

$$\triangleq F_k \begin{bmatrix} x_k \\ \theta_k \end{bmatrix} \tag{6.15}$$

Next we choose a Lyapunov function for (6.15) given by $V = x_k^T P x_k + \epsilon^{-1}\theta_k^T\theta_k$ where P is a real positive definite symmetric matrix satisfying the Popov-Kalman-Yakubovich equations of Lemma 2.10 with SPR transfer function $\bar{d}+\bar{c}^T(zI-\bar{A})^{-1}\bar{b}$.

Obviously V is positive definite and direct substitution of $\bar{P} = P \oplus \epsilon^{-1}I$ and $\bar{F}_k$ into the Lyapunov equation (6.11) admits the solution

$$N_k = \begin{bmatrix} \gamma I & L + w(\delta+\epsilon\phi_k^T\phi_k)\chi_k\bar{c} & \xi_k\chi_k\bar{c} \\ 0 & -\chi_k w\phi_k & \xi_k\chi_k\bar{d}\phi_k \end{bmatrix}$$

where $\chi_k = (1+\bar{d}\epsilon\phi_k^T\phi_k+\bar{d}\delta)^{-1}$ and $\xi_k = (2\delta+\epsilon\phi_k^T\phi_k)^{1/2}$. Thus (6.15) is stable. Now choose

$$K_k = \begin{bmatrix} 0 & -w^{-1}\bar{b} & 0 \\ 0 & 0 & -\epsilon\xi_k^{-1}\phi_k \end{bmatrix}$$

where K_k is bounded because of our device of choosing $\delta > 0$. This results in

$$\bar{F}_k - K_k N_k^T = \begin{bmatrix} A + w^{-1}\bar{b}\bar{L} & 0 \\ 0 & I \end{bmatrix}$$

and we write

$$N_k = \begin{bmatrix} \gamma I & \bar{c} & 0 \\ 0 & 0 & \phi_k \end{bmatrix} \begin{bmatrix} I & \gamma^{-1}L & 0 \\ 0 & w(\delta+\epsilon\phi_k^T\phi_k)\chi_k I & \xi_k\chi_k I \\ 0 & -w\chi_k I & \xi_k\chi_k\bar{d}I \end{bmatrix}$$

The second matrix above, R_k, is bounded and has bounded inverse (again a consequence of $\delta > 0$) and it is easily seen that uniform complete observability of $[\bar{F}_k - K_k N_k^T, N_k]$ is equivalent to the same property for $[\bar{F}_k - K_k N_k^T, N_k R_k^{-1}]$. Further the block diagonal structure

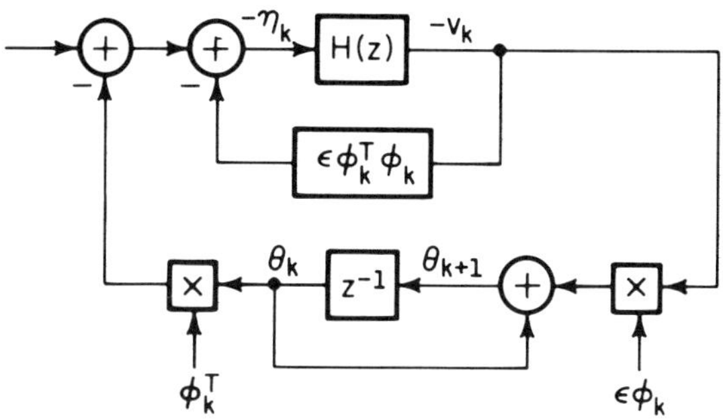

Fig. 2.4 Discrete-time adaptive error system.

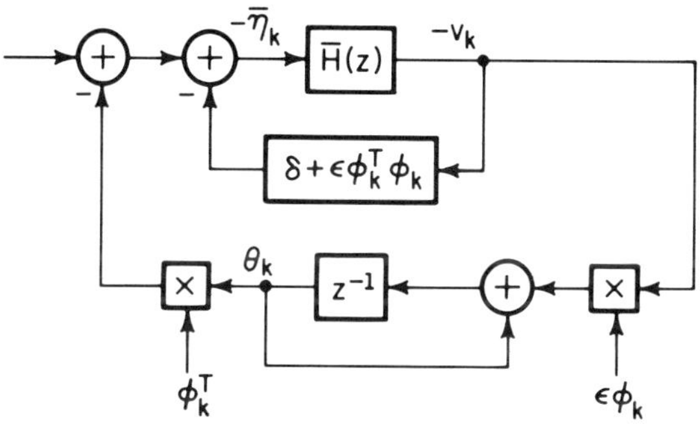

Fig. 2.5 Redrawing of discrete-time error system.

of $\bar{F}_k - K_k N_k^T$ and $N_k R_k^{-1}$ means that this latter pair is uniformly completely observable if and only if both $[\bar{A}+w^{-1}\bar{b}L, \gamma I]$ and $[I, \phi_k]$ are uniformly completely observable. The first pair is trivially observable while the second pair has the property if ϕ_k satisfies (6.14) for some T, α and β. Thus (6.15) is exponentially stable. The exponential stability of (6.9) is then immediate.

The final claim of the theorem statement will be more simply verified with the averaging techniques of the next chapter.

□

The remarks of the previous sections carry over directly to this result which describes sufficient conditions for the exponential stability of the adaptive system error equations. The theorem has been deliberately stated to parallel Theorem 2.3 although, as was noted in the proof, the SPR condition dictates via (6.13) that $H(\infty) = d > 0$ so that consideration of the strictly proper case is obviated and the problem of dealing with increasingly rapidly varying signals does not arise in discrete-time. Otherwise, we still have global stablity subject to the SPR condition and global exponential stability with the addition of a persistence of excitation requirement.

2.6.3 Relaxing the Discrete SPR Condition

Again to parallel our earlier development we consider relaxing the SPR condition of $H(z)$ at the expense of restricting the frequency content of $\{\phi_k\}$. Starting from the postulate that at low frequencies, i.e., near $z = 1$, $\mathrm{Re}H(e^{j\omega}) > 0$, and that $H(z)$ is stable, we assume that $H(z)$ may be written as

$$H(z) = H_1(z) + (1 - z^{-1})H_2(z)$$

where $H_1(z)$ is SPR and $H_2(z)$ is strictly stable. We may then write (1.2) as

$$\begin{aligned}\theta_{k+1} &= \theta_k - \epsilon\phi_k H_1(z)\{\phi_k^T\theta_{k+1}\} \\ &\quad - \epsilon\phi_k H_2(z)\{[\phi_k^T - \phi_{k-1}^T]\theta_{k+1}\} \\ &\quad - \epsilon\phi_k H_2(z)\{\phi_k^T(1-z^{-1})\theta_{k+1}\} \qquad (6.16)\end{aligned}$$

The successive terms in (6.16) are analyzed in turn.

From Theorem 2.9 we know that, subject to $\{\phi_k\}$ satisfying (6.14) and with $h_k = 0$,

$$\theta_{k+1} = \theta_k - \epsilon\phi_k H_1(z)\{\phi_k^T\theta_{k+1}\} + h_k \tag{6.17}$$

is exponentially stable. Consequently, from Lemma 2.1 (iii') the l_p gain for $p \in [1,\infty)$ from h_k to θ_{k+1} is m/λ for some $m > 0$ and $0 < \lambda < 1$. Now consider the system described by (6.17) interconnected with

$$q_k = \epsilon\phi_k H_2(z)\{[\phi_k^T - \phi_{k-1}^T]\theta_{k+1}\} \tag{6.18}$$

$$h_k = -q_k$$

By the small gain theorem, provided the gain of the operator from θ_{k+1} to q_k is smaller than λ/m, the interconnection will be stable. Assume that

$$\|\phi_k\|_\infty = m_1 < \infty \tag{6.19}$$

$$\|\phi_k - \phi_{k-1}\|_\infty = m_2\ (\leq 2m_1) < \infty \tag{6.20}$$

and denote the l_p gain of $H_2(z)$ by m_3; then the interconnection (6.17)-(6.18) will be stable provided

$$\epsilon\frac{m}{\lambda}m_1m_2m_3 < 1 \tag{6.21}$$

Exponential stability requirements can be derived by exponential weighting as before in Theorem 2.5. (This requires that we consider signals like $\theta_{a,k} = a^k\theta_k$ for some $a > 1$, and proceed as earlier.)

The final term on the right of (6.16) is included as a regular perturbation to the exponentially stable interconnection immediately above. Denote the operator $\epsilon\phi_k H_2(z)\{\phi_k^T(1-z^{-1})\}$, by N; then in our previous notation the l_p gain of N is bounded as follows:

$$\|N\|_p \leq \epsilon m_1m_3m_1 2 = 2\epsilon m_1^2m_3$$

and we may write (6.16) as

$$(I + \epsilon N)\theta_{k+1} = -\epsilon M\theta_{k+1} \tag{6.22}$$

where M is the operator including only the first three terms on the right hand side of (6.16). Once again, provided

$$\epsilon\|N\| < 1 \tag{6.23}$$

we may invert $(I+\epsilon N)$ as

$$(I+\epsilon N)^{-1} = I + K$$

where

$$K = -\epsilon N + \epsilon^2 N^2 - \epsilon^3 N^3 + \cdots$$

and, hence,

$$\|K\| \leq \epsilon\|N\|(1 - \epsilon\|N\|)^{-1} \tag{6.24}$$

From (6.22) we see that we may write (6.16) as

$$\theta_{k+1} = -\epsilon M \theta_{k+1} - \epsilon K M \theta_{k+1}$$

where $\theta_{k+1} = -\epsilon M \theta_{k+1} + h_k$ is exponentially stable with gain from h_k to θ_{k+1} of m'/λ'. Provided

$$\frac{\epsilon m'}{\lambda'}\|K\|\,\|M\| < 1 \tag{6.25}$$

(exponential) stability of (6.16) follows.

The assumptions leading to these sufficient conditions for exponential stability of (6.16) are grouped in the following theorem.

Theorem 2.10:

Consider the difference equation (1.2)

$$\theta_{k+1} = \theta_k - \epsilon\phi_k H(z)\{\phi_k^T \theta_{k+1}\}$$

Suppose $H(z)$ is a stable discrete transfer function which may be written

$$H(z) = H_1(z) + (1-z^{-1})H_2(z)$$

where $H_1(z)$ is discrete SPR, and $H_2(z)$ is strictly stable and let $0 < \rho < 1$ be such that $H_1(\rho z)$ and $H_2(\rho z)$ have no poles in $|z| > 1$. Further suppose that $\{\phi_k\}$ is persistently exciting and satisfies

$$\|\phi_k\|_\infty = m_1 < \infty$$

and

$$\|\phi_k - \phi_{k-1}\|_\infty = m_2$$

Then the difference equation will be exponentially stable if all the following conditions hold:

$$\text{(i)} \quad \epsilon m_1 m_2 \gamma_2(\rho) \frac{m}{\rho\lambda} < 1 \tag{6.26}$$

$$\text{(ii)} \quad 2\epsilon m_1^2 \gamma_2(\rho) < 1 \tag{6.27}$$

$$\text{(iii)} \quad \epsilon^2 \frac{m'}{\rho\lambda'} \frac{m_1^2\gamma_2(\rho)}{1-\epsilon m_1^2\gamma_2(\rho)} \cdot [m_1^2\gamma_1(\rho) + m_1 m_2 \gamma_2(\rho)] < 1 \tag{6.28}$$

where:

$$\gamma_i(\rho) = \sup_{\theta \in [0,2\pi]} |H_i(\rho e^{j\theta})|$$

$i = 1,2;$ m and λ are associated with the convergence rate of (6.17) and m', λ' with (6.17)-(6.18).

Proof

Explicit evaluation of inequalities (6.21), (6.23) and (6.25) yields the respective conditions (i) - (ii).

□

The similarity between this result and Theorem 2.6 is immediate and, consequently, the same broad conclusions hold. It is worth noting that while (6.19) yields $\|\phi_k - \phi_{k-1}\| \le 2m_1$ by the triangle inequality, we have chosen to assume different m_2 in (6.20) since in line with our earlier analysis, we expect the frequency content of the regressor signals, which can be estimated by m_2/m_1 to be important. Indeed, as before, inequalities (6.27) and (6.28) are able to be satisfied by taking ϵ sufficiently small while (6.26) remains the active constraint.

From Theorem 2.9 we know that m/λ scales as $(m_1^2\epsilon)^{-1}$ so that, once ϵ is chosen small enough, the parameter m_2/m_1 becomes the sole design variable here. That is, slow adaptation together with maintenance of the dominant frequency content of $\{\phi_k\}$ where $H(z)$ is roughly SPR is sufficient for exponential stability of the adaptive error system. We reiterate here that stability above is conditional on persistence of excitation of $\{\phi_k\}$ and that it may be a nontrivial problem to ensure persistence of excitation by manipulating available signals while still ensuring the dominant frequency requirements on the whole ϕ_k−vector. In later chapters these competing requirements for exponential stability of the homogeneous algorithm will be compounded by further restrictions on boundedness of additive driving signals and sensitivity functions. To a very real extent the ability to satisfy simultaneously this class of conflicting constraints dictates the viability of adaptive control for particular plants.

2.7 NOTES AND REFERENCES

The explicit appearance of the SPR condition in the stability of adaptive systems has a long history usually being traced back to Butchart and Shackloth (1965) and Parks (1966), and more recently in Landau (1979) and Narendra, Lin, and Valavani (1980). The joint use of persistence of excitation and SPR to achieve exponential stability appears in Anderson (1977), Morgan and Narendra (1977), and Yuan and Wonham (1977) in various guises in continuous-time. The Lyapunov development in this chapter is derived from Anderson (1977). In discrete time, the Lyapunov theory is established by Bitmead and Anderson (1980a) and is applied to derive exponential stability subject to persistence of excitation and SPR in Bitmead and Anderson (1980b), Anderson and Johnson (1982), and Goodwin and Sin (1984). References on the Popov-Kalman-Yakubovich Lemma, positive real functions, and passivity are Anderson and Vongpanitlerd (1973) and Narendra and Taylor (1973). The small gain theorem is from Zames (1966) and Desoer and Vidyasagar (1975).

The techniques of this chapter for relaxing the SPR condition were first given in Anderson et al. (1984) for both continuous- and discrete-time. The shifting of persistence of excitation conditions, from the full regression vector to richness of input signal conditions only, has been presented by Anderson and Johnson (1982a) for discrete-time and our derivation of continuous-time results follows Mareels (1984) with the connection to output reachability following Green and Moore (1985). These conditions on richness of input signals can be nicely interpreted in terms of sinusoidal frequency components or spectral lines as is expounded by Boyd and Sastry (1983).

Chapter 3

TIME-SCALE DECOMPOSITION AND AVERAGING ANALYSIS

3.1 INTRODUCTION

The small gain analysis of the preceding chapter has shown that the SPR condition for stability can be relaxed if the class of signals ϕ is restricted by imposing some bounds on $\|\phi\|$ and $\|\dot{\phi}\|$. In this chapter we further explore the dependence of stability properties of the adaptation loop on characteristics of the signal ϕ. The main result of this chapter is a stability criterion which delineates a sharp boundary between the signals ϕ for which the adaptation loop with a non-SPR transfer function H is stable and those for which the adaptation is unstable. In the case of $T-$periodic[‡] signals the criterion examines the eigenvalues of the matrix

$$\phi H(\phi^T) \tag{1.1}$$

averaged over the period T. For a more general class of signals, the stability depends on properties of sample averages of the matrix

$$\phi H(\phi^T) + [\phi H(\phi^T)]^T \tag{1.2}$$

The stability criteria are derived using a time-scale decomposition of the adaptation loop. The convolution operator form

---

[‡] Although the same symbol T is used to denote periodicity, transposition, and truncation, its specific meaning will be clear from the context.

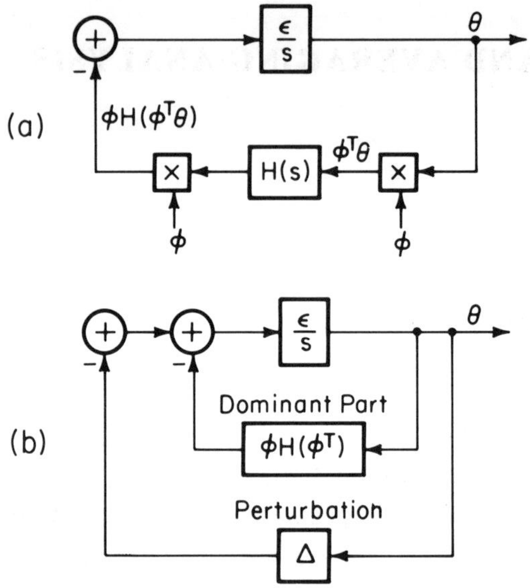

Fig. 3.1 Operator decomposition of the linearized adaptive system (a) into its dominant part $\phi H(\phi^T)$ and Δ-perturbation (b).

of this decomposition is presented in Section 3.2 and its state-space form in Section 3.3. In contrast to the transfer function decomposition into an SPR part and a non-SPR part used in the preceding chapter, the time-scale decomposition retains the original non-SPR transfer function H in Figure 3.1a, unchanged. Instead, it decomposes the feedback signal as shown in Figure 3.1b. Of the two feedback paths, the inner path $\phi H(\phi^T)$ is dominant, while the outer path is a perturbation which is small when the adaptation is slow, that is, when ϵ is small.

The stability properties of the inner loop, analyzed in Section 3.4, clearly show which restrictions should be imposed on the spectral content of ϕ when H is not SPR. These restrictions are incorporated in the stability criteria formulated in Section 3.5. The *average SPR criterion* has an extremely simple frequency domain interpretation: the signals ϕ are restricted to a class in which the non-SPR transfer function H behaves, *on average*, as an SPR transfer function.

A further implication of the stability criteria is that, if the class of signals ϕ is not restricted, then the SPR property of H is not only a sufficient, but also a necessary condition for stability. In other words, *if H is not SPR, there always exist signals for which the linear adaptation loop is unstable*. Examples illustrating these issues are discussed in Sections 3.6 and 3.7.

3.2 DECOMPOSITION OF THE CONVOLUTION OPERATOR

Our goal in this and the next section is to demonstrate that the feedback path with the matrix $\phi H(\phi^T)$ is the dominant part of the slow adaptation system in the Figure 3.1a.

3.2.1 Operator Decomposition (Continuous Time)

In continuous time, the slow adaptation system is described by

$$\dot{\theta} = -\epsilon\phi(t)V(\theta)(t) \tag{2.1}$$

where the action of the transfer function H upon the scalar input $\phi^T(t)\theta(t)$, namely $H(\phi^T\theta)$, is represented by the operator

$$V(\theta)(t) = \int_{-\infty}^{t} h(t-\tau)\phi^T(\tau)\theta(\tau)d\tau + d\phi^T(t)\theta(t) \tag{2.2}$$

The following decomposition result holds:

Lemma 3.1:

Under the assumption that the kernel $h(t)$ is exponentially stable:

$$|h(t-\tau)| \leq K_h e^{a(t-\tau)} \quad , \quad \forall\, t > \tau \tag{2.3}$$

for some positive constants K_h and a and that the regressor vector $\phi(t)$ is a uniformly bounded integrable function of t,

$$\|\phi\|_\infty = \Phi < \infty \tag{2.4}$$

the operator

$$\phi(t)V(\theta)(t) = \phi(t)[\int_{-\infty}^{t} h(t-\tau)\phi(\tau)^T\theta(\tau)d\tau + d\phi(t)^T\theta(t)] \tag{2.5}$$

can be approximated by

$$\phi(t)v_0^T(t)\theta(t) = \phi(t)[\int_{-\infty}^{t} h(t-\tau)\phi(\tau)^T d\tau + d\phi(t)^T]\theta(t) \tag{2.6}$$

The error, denoted by $\Delta(\theta)$,

$$\Delta(\theta) = \phi(t)(V(\theta)(t) - v_0^T(t)\theta(t))$$

is of the order of $\dot{\theta}(t)$:

$$\|\Delta(\theta)\|_T \leq \Phi^2 \frac{K_h}{a^2} \|\dot{\theta}\|_T$$

Corollary 3.1:

Applying this result to the system (2.1), one obtains that the operator $\Delta(\theta)$ has a gain proportional to ϵ,

$$\|\Delta(\theta)\|_T < \epsilon\Phi^4 \frac{K_h}{a^2}\left(\frac{K_h}{a} + d\right)\|\theta\|_T \tag{2.7}$$

where subscript T denotes truncation and $\|\cdot\|$ denotes supremum norm, as defined in Chapter 2.

Proof

Integrating (2.2) by parts we obtain:

$$\begin{aligned} V(\theta)(t) &= \int_{-\infty}^{t} h(t-\tau)\phi^T(\tau)\theta(\tau)d\tau + d\phi^T(t)\theta(t) \\ &= \{\int_{-\infty}^{t} h(t-\tau)\phi^T(\tau)d\tau + d\phi^T(t)\}\theta(t) \\ &\quad - \int_{-\infty}^{t}(\int_{-\infty}^{\tau} h(t-s)\phi^T(s)ds)\dot{\theta}(\tau)d\tau \end{aligned}$$

Hence

$$\|\Delta(\theta)\|_T \leq \sup_{t\leq T} \int_{-\infty}^{t}\int_{-\infty}^{\tau} K_h e^{-a(t-s)} ds d\tau \Phi^2 \|\dot{\theta}\|_T \leq K_h \frac{\Phi^2}{a^2}\|\dot{\theta}\|_T$$

This proves the Lemma. From (2.1) and (2.2) we obtain an estimate for $\|\dot{\theta}\|_T$:

$$\|\dot{\theta}\|_T \leq \epsilon\Phi^2\left(\frac{K_h}{a} + d\right)\|\theta\|_T$$

Combining this estimate with the bound for $\Delta(\theta)$ proves the corollary.

□

We have, therefore, demonstrated that the decomposition in Fig 3.1b is valid, that is the inner loop described by

$$\dot{\theta} = -\epsilon\phi(t)v_0^T(t)\theta \tag{2.8}$$

is dominant, while $\Delta(\theta)$ is an $O(\epsilon)$−perturbation.

The result of Lemma 3.1 expresses the time-scale property that, due to the smallness of ϵ, the variations of $\theta(t)$ are slow. This time-scale property allows $H(\phi^T\theta)$ to be approximated by $H(\phi^T)\theta$.

3.2.2 Operator Decomposition (Discrete Time)

The discrete-time counterpart of (2.1), (2.2) is

$$\theta(k+1) = \theta(k)-\epsilon h(0)\gamma(k)\phi^T(k)\theta(k+1)-\epsilon\gamma(k)V(\theta)(k) \quad (2.9)$$

$$V(\theta)(k) = \sum_{l=-\infty}^{k} h(k-l+1)\phi^T(l-1)\theta(l) \quad (2.10)$$

where $\gamma(k)$ depends on the particular discrete time adaptive algorithm being used. When $\gamma(k) \equiv \phi(k)$, which is the case for some algorithms (see Chapter 5), the analogy with the continuous time case (2.1) is complete. The following result is the discrete counterpart of Lemma 3.1:

Lemma 3.2:

Under the assumption that the kernel $h(k)$ is exponentially stable, i.e.,

$$|h(k-l)| \leq K_h a^{k-l} \qquad \forall\ k \geq l \quad (2.11)$$

for some positive constants $a < 1$ and K_h; and that the regressor $\phi(k)$ is uniformly bounded

$$\|\phi\|_\infty = \Phi < \infty \quad (2.12)$$

the operator

$$V(\theta)(k) = \sum_{l=-\infty}^{k} h(k-l+1)\phi^T(l-1)\theta(l) \quad (2.10)$$

can be approximated by

$$v^T(k)\theta(k) = \sum_{l=-\infty}^{k} h(k-l+1)\phi^T(l-1)\theta(k) \quad (2.13)$$

The error is of the order of $\theta(k+1)-\theta(k) = \tilde{\theta}(k+1)$:

$$\|V(\theta)(\cdot)-v^T(\cdot)\theta(\cdot)\|_T < \Phi K_h \frac{a^2}{(1-a)^2}\|\tilde{\theta}\|_T \quad (2.14)$$

Proof

$$V(\theta)(k)-v^T(k)\theta(k) = \sum_{l=-\infty}^{k-1} h(k-l+1)\phi^T(l-1)(\theta(l)-\theta(k))$$

$$= \sum_{l=-\infty}^{k-1} h(k-l+1)\phi^T(l-1)\sum_{m=0}^{k-l-1}(\theta(l+m)-\theta(l+1+m))$$

Hence, the result follows from

$$\|V(\theta)(\cdot)-v^T(\cdot)\theta(\cdot)\|_T \le \Phi \sup_{k\le T}\left(\sum_{l=-\infty}^{k-1}\sum_{m=0}^{k-l-1} K_h a^{k-l+1}(k-l)\right)\|\tilde{\theta}\|_T$$

$$\le \Phi K_h \frac{a^2}{(1-a)^2}\|\tilde{\theta}\|_T$$

□

Corollary 3.2:

Applying this result to (2.9) under the additional assumption that the gain sequence $\gamma(k)$ is bounded

$$\|\gamma\|_\infty = \Gamma \tag{2.15}$$

we obtain:

$$\|V(\theta)(\cdot)-v^T(\cdot)\theta(\cdot)\|_T \le \epsilon\frac{K_h^2\Phi^2\Gamma a^2}{(1-a)^3}\|\theta\|_T \tag{2.16}$$

Proof

This follows readily from (2.9): bounding $\tilde{\theta}_k$ as

$$\|\tilde{\theta}(\cdot)\|_T \le \epsilon|h(0)|\Gamma\Phi\|\theta\|_T + \epsilon\Gamma\frac{aK_h}{1-a}\Phi\|\theta\|_T$$

$$\le \epsilon\Gamma K_h\frac{1}{1-a}\Phi\|\theta\|_T$$

and substituting this bound for $\|\tilde{\theta}\|_T$ into (2.14) yields (2.16), because $|h(0)|\le K_h$ in view of (2.11).

□

Remark: Assuming (2.15) is equivalent to requiring that the inverse

$$(I+\epsilon h(0)\gamma(k)\phi^T(k))^{-1} = I-\epsilon g(k,\epsilon)h(0)\gamma(k)\phi^T(k) \tag{2.17}$$

exists for all k, where

$$g(k,\epsilon) = \frac{1}{1+\epsilon h(0)\gamma^T(k)\phi(k)} \tag{2.18}$$

A sufficient condition for (2.17) to be well defined is

$$\epsilon|h(0)|\Gamma\Phi < 1 \tag{2.19}$$

In view of this Lemma, (2.9) can be rewritten as

$$\theta(k+1) = \theta(k)-\epsilon[h(0)\gamma(k)\phi^T(k)\theta(k+1)-\epsilon v^T(k)\theta(k)+\Delta(\theta)] \tag{2.20}$$

where Δ is an operator with gain proportional to ϵ. Consequently, using the Lemma once more to replace $\theta(k+1)$ by $\theta(k)$ in the right hand side of (2.20), we obtain for the dominant part of (2.9)

$$\theta(k+1) = [I-\epsilon\gamma(k)v_0^T(k)]\theta(k) \tag{2.21}$$

where v_0^T is defined by

$$v_0^T(k) = h(0)\phi(k)+\sum_{l=-\infty}^{k} h(k-l+1)\phi^T(l-1) \tag{2.22}$$

which is precisely $(H\phi^T)(k)$ and describes the discrete counterpart of the inner loop in Figure 3.1b. The analogy with the continuous-time case is complete.

3.3 STATE-SPACE DECOMPOSITION

The developments thus far have not been restricted to finite-dimensional systems. Let us now assume that the transfer function H has a finite-dimensional, say m-dimensional, representation

$$H(s) = d + c^T(sI-A)^{-1}b \tag{3.1}$$

and that the $m\times m$ matrix A is Hurwitz,

$$\mathrm{Re}\lambda(A) < 0 \tag{3.2}$$

In view of this assumption, there exist positive constants a and $K \geq 1$ such that

$$|e^{A(t-\tau)}| < Ke^{-a(t-\tau)} \ , \quad \forall\ t > \tau \tag{3.3}$$

The state-space form of the adaptation equation (2.1) with operator V defined in (2.2) is then

$$\begin{bmatrix} \dot{\theta} \\ \dot{x} \end{bmatrix} = \begin{bmatrix} -\epsilon d\phi(t)\phi^T(t) & -\epsilon\phi(t)c^T \\ b\phi^T(t) & A \end{bmatrix} \begin{bmatrix} \theta \\ x \end{bmatrix} \tag{3.4}$$

$$\theta \in \mathbb{R}^n \ , \quad x \in \mathbb{R}^m \tag{3.5}$$

This linear time-varying system is a two-time-scale system because, when ϵ is small, $\dot{\theta}$ is much smaller than $\dot{x}$, and, hence, $\theta(t)$ is slower than $x(t)$.

3.3.1 L-Transformation (Continuous Time)

To separate the slow and fast parts of (3.4) we introduce the transformation ("L-Transformation")

$$\begin{bmatrix} \theta \\ x \end{bmatrix} = \begin{bmatrix} I_n & 0 \\ -L(t,\epsilon) & I_m \end{bmatrix} \begin{bmatrix} \theta \\ x \end{bmatrix} \tag{3.6}$$

and require that the $m\times n$ matrix $L(t,\epsilon)$ be a solution of

$$\dot{L} = AL + b\phi^T(t) + \epsilon dL\phi(t)\phi^T(t) + \epsilon L\phi(t)c^TL \tag{3.7}$$

In view of (3.7), the substitution of (3.6) into (3.4) results in

$$\begin{bmatrix} \dot{\theta} \\ \dot{z} \end{bmatrix} = \begin{bmatrix} -\epsilon\phi(t)[d\phi^T(t)+c^TL(t,\epsilon)] & -\epsilon\phi(t)c^T \\ 0 & A+\epsilon L(t,\epsilon)\phi(t)c^T \end{bmatrix} \begin{bmatrix} \theta \\ z \end{bmatrix} \tag{3.8}$$

If such a matrix $L(t,\epsilon)$ exists, the time-scale decomposition is achieved, and the stability properties of the fast $z-$system

$$\dot{z} = (A + \epsilon L(t,\epsilon)\phi(t)c^T)z \tag{3.9}$$

and of the slow $\theta-$system

$$\dot{\theta} = - \epsilon\phi(t)[d\phi^T(t) + c^TL(t,\epsilon)]\theta \tag{3.10}$$

can be separately analyzed. The forcing term $-\epsilon\phi(t)c^Tz$ is not included in (3.10), because it does not influence the stability properties of this linear system. Notice that (if L is a bounded matrix function of t) the original system (3.4) has the same stability properties as the transformed system (3.8) because the transformation preserves the stability properties. The stability analysis of (3.9) and (3.10) is undertaken in subsequent sections, and its conclusions carry immediately over to the original system (3.4). In this section we interpret the meaning of $L(t,\epsilon)$ and formulate a sufficient condition for its existence.

We do not specify an initial condition for the $L-$equation (3.7) directly. Instead, we assume that the signal $\phi(t)$ is defined for all $t \in (-\infty,\infty)$, as in (2.4). Then, in view of (3.2), the matrix

$$L_0(t) = \int_{-\infty}^{t} e^{A(t-\tau)}b\phi^T(\tau)d\tau \tag{3.11}$$

specifies $L(t,0) = L_0(t)$ as the solution $L(t,\epsilon)$ of (3.7) for $\epsilon = 0$ with the initial condition $L(0,0) = L_0(0)$. By a standard theorem on differentiability with respect to parameters, the derivative $\frac{dL(t,\epsilon)}{d\epsilon}$ exists at $\epsilon = 0$, and hence

$$L(t,\epsilon) = L_0(t) + \epsilon L_1(t,\epsilon) \tag{3.12}$$

where $L_1(t,\epsilon)$ is continuous and bounded near $\epsilon = 0$. This fact proves the existence of $L(t,\epsilon)$ and its continuity with respect to t and ϵ near $\epsilon = 0$. A complete proof and a bound for ϵ is given in Lemma 3.3.

Let us first introduce the notation

$$v^T(t,\epsilon) = d\phi^T(t) + c^T L(t,\epsilon) = d\phi^T(t) + c^T L_0(t) + O(\epsilon) \qquad (3.13)$$

where the signal

$$v_0^T(t) = v^T(t,0) = d\phi^T(t) + c^T L_0(t) \qquad (3.14)$$

has already been defined in (2.6). It is clear from (3.13) that the dominant part (i.e., disregarding $O(\epsilon^2)$ terms) of the system (3.10) is

$$\dot{\theta} = -\epsilon\phi(t)v_0^T(t)\theta \qquad (3.15)$$

which is the same as the system (2.8), obtained by the operator decomposition.

In this way, the original adaptation loop in Figure 3.2a has been transformed into the scheme shown in Fig 3.2b. Its main part is the θ-loop described by (3.15), with the input $\phi c^T z$ and the coefficient matrix ϕv^T formed in the $z-$system and the $L-$system, respectively, neither of which depends on θ.

3.3.2 Existence of L (Continuous Time)

We proceed to derive a condition under which a unique $L(t,\epsilon)$ exists.

Lemma 3.3:

Assuming that the $m \times m$ matrix A is Hurwitz and that the regressor $\phi(t) \in \mathbb{R}^n$ is a bounded (condition (2.12)) integrable function of t, there exist ϵ^* and l^* such that for all $\epsilon \in (-\epsilon^*, \epsilon^*)$ there exists a unique, continuous, and bounded $m \times n$ matrix function $L(t,\epsilon)$ satisfying:

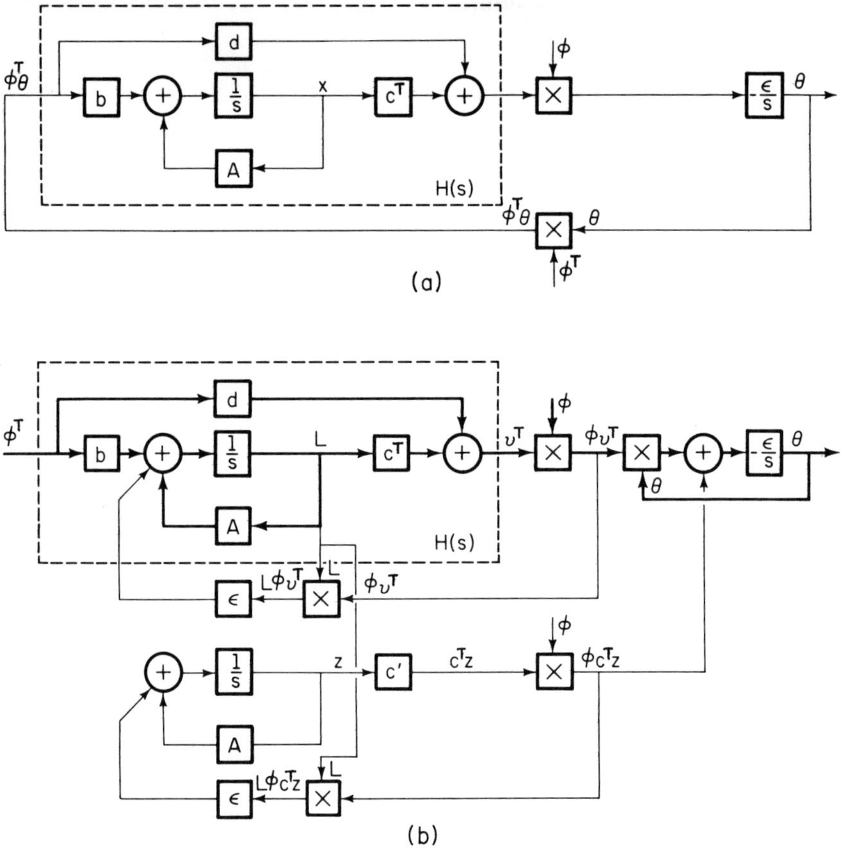

Fig. 3.2 The linearized feedback loop (a) transformed into the series connection (b). Note the transfer function *H(s)* in the *L*-system.

$$\dot{L} = AL + b\phi^T(t) + \epsilon d\phi^T(t) + \epsilon L\phi(t)c^T L \qquad (3.7)$$

$$L(t,\epsilon) = L_0(t) + \epsilon L_1(t,\epsilon) \qquad (3.12)$$

where $L_0(t)$ is the steady state solution (3.11) of Equation (3.7) for $\epsilon = 0$. $L(t,\epsilon)$ is bounded :

$$\|L(\cdot,\epsilon)\| \leq (1 + |\epsilon| l^*) \|L_0\| \tag{3.16}$$

$$\|L_1(\cdot,\epsilon)\| \leq l^* \|L_0\| \tag{3.17}$$

This matrix $L(t,\epsilon)$ used in the transformation (3.6) decomposes subsystem (3.4) into a slow (3.10) and a fast subsystem (3.9), which determine the stability properties of the original system (3.4).

□

Remark: Bounds ϵ^*, l^* are found by optimizing the allowable range of ϵ over the inequalities:

$$\epsilon^2 l^2 \left(|c| \|L_0\| \Phi \frac{K}{a} \right) + |\epsilon| l \Phi \frac{K}{a} (|d|\Phi + 2|c| \|L_0\|)$$

$$+ \frac{K}{a} \Phi(|d|\Phi + |c| \|L_0\|) - l \leq 0 \tag{3.18}$$

$$\epsilon^2 l \left(2|c| \|L_0\| \Phi \frac{K}{a} \right) + |\epsilon| \Phi \frac{K}{a} (|d|\Phi + 2|c| \|L_0\|) - 1 < 0 \tag{3.19}$$

Although an explicit bound for ϵ can be obtained by a maximization with respect to l, subject to constraints (3.18) and (3.19), we postpone this operation until the stability analysis of the system (3.8) is completed in Section 3.5, because the stability considerations will introduce further constraints to be met by ϵ and l. However, an immediate observation important for the stability analysis is that (3.18) implies

$$l \geq \frac{K}{a} \Phi(|d|\Phi + |c| \|L_0\|) > 0 \tag{3.20}$$

Furthermore, (3.18) guarantees that the z-system (3.9) is exponentially stable, because (3.18) is more restrictive than

$$\epsilon^2 l \left(|c| \|L_0\| \Phi \frac{K}{a} \right) + |\epsilon| \Phi \frac{K}{a} |c| \|L_0\| - 1 < 0 \tag{3.21}$$

This inequality follows from the application of the Total Stability Theorem (Theorem 1.3, Chapter 1) to (3.9), using (3.16) and (3.3). We will return to this issue in Section 3.5.

Proof

In the Banach space $C(-\infty,\infty)$ of continuous, bounded $m \times n$ matrix functions of t, equipped with the supremum norm, consider the closed subset

$$C_l := \{Y \in C: \|Y\| \leq (1+|\epsilon|l)\|L_0\|\} \ , \ l > 0 \tag{3.22}$$

Note that, in view of (3.3), (2.12) and (3.11)

$$\|L_0\| \leq |b|\Phi \frac{K}{a} \tag{3.23}$$

Define, on the subset C_l, the operator M:

$$M(Y)(t) = \epsilon \int_{-\infty}^{t} e^{A(t-\tau)} G(\tau, Y(\tau)) d\tau + L_0(t) \tag{3.24}$$

where G is defined as

$$G(t,Y) = dY\phi(t)\phi^T(t) + Y\phi(t)c^T Y \tag{3.25}$$

From the inequality

$$\|M(Y)\| \leq \|L_0\| \left(1+|\epsilon|\frac{K}{a}(|d|(1+|\epsilon|l\Phi^2)+(1+|\epsilon|l)^2\Phi|c|\,\|L_0\|) \right) \tag{3.26}$$

it follows that M is well defined on C_l for all ϵ, l satisfying (3.18). From the inequality

$$\|M(Y_1)-M(Y_2)\| \leq |\epsilon|\frac{K}{a}(|d|\Phi^2+2(1+|\epsilon|l)\Phi|c|\,\|L_0\|)\,\|Y_1-Y_2\|$$

it follows that M is a contraction operator on C_l for all ϵ, l satisfying (3.19).

Hence, M has a unique fixed point $L(t,\epsilon)$. The bounds (3.16) and (3.17) follow directly from the definition (3.12), the construction (3.22) of C_l, and the inequality (3.26).

□

3.3.3 L-Transformation (Discrete Time)

In the discrete-time case, when the transfer function H has an $m-$dimensional state representation analogous to (3.1), we rewrite (2.9)-(2.10) as

$$\theta(k+1) = \theta(k)-\epsilon\gamma(k)[d\phi^T(k)\theta(k+1)+c^Tx(k)]\ ,\ \theta \in \mathbb{R}^n \tag{3.27}$$

$$x(k+1) = Ax(k)+b\phi^T(k)\theta(k+1)\ ,\ x \in \mathbb{R}^m \tag{3.28}$$

and, analogously to (3.2) - (3.3), assume that

$$|\lambda(A)| < 1\ ,\ |A^k| \leq Ka^k\ ,\ \forall\ k \tag{3.29}$$

for some positive constants $a < 1$ and $K \geq 1$. Assuming that (cf. (2.19))

$$|\epsilon||d|\Phi\Gamma < 1 \tag{3.30}$$

so as to guarantee the existence of the inverse (2.17), we rewrite (3.27) - (3.28) in the explicit form of equation (6.9) in Chapter 2, namely

$$\begin{bmatrix}\theta(k+1)\\ x(k+1)\end{bmatrix} = \begin{bmatrix}I-\epsilon g(k,\epsilon)d\gamma(k)\phi^T(k) & -\epsilon g(k,\epsilon)\gamma(k)c^T\\ bg(k,\epsilon)\phi^T(k) & A-\epsilon g(k,\epsilon)\phi^T(k)\gamma(k)bc^T\end{bmatrix}\begin{bmatrix}\theta(k)\\ x(k)\end{bmatrix} \tag{3.31}$$

where $g(k,\epsilon)$ is defined as in (2.18) with $d = h(0)$. The time-scale decomposition of (3.31) is accomplished using the transformation

$$\begin{bmatrix}\theta(k)\\ z(k)\end{bmatrix} = \begin{bmatrix}I_n & 0\\ -L(k,\epsilon) & I_m\end{bmatrix}\begin{bmatrix}\theta(k)\\ x(k)\end{bmatrix} \tag{3.32}$$

where $L(k,\epsilon)$ is this solution of the difference equation

$$\begin{aligned} L(k+1,\epsilon) &= AL(k,\epsilon) + b\phi^T(k) + \epsilon[L(k+1,\epsilon)-b\phi^T(k)]\\ &\quad \cdot g(k,\epsilon)\gamma(k)[\phi^T(k)d+c^TL(k,\epsilon)] \end{aligned} \tag{3.33}$$

which, as $\epsilon \to 0$, converges to $L_0(k)$ defined by

$$L_0(k) = \sum_{j=-\infty}^{k} A^{k-j} b \phi^T(j) \tag{3.34}$$

We note that $L_0(k)$ is the steady state solution of the difference equation (3.33) for $\epsilon = 0$:

$$L_0(k+1) = AL_0(k) + b\phi(k) \tag{3.35}$$

In view of (3.29) and (2.12), $L_0(k)$ is bounded by

$$\|L_0\| \leq K \frac{\Phi}{1-a} |b| \tag{3.36}$$

Clearly, $L_0(k)$ is by construction a solution $L(k,0)$ of (3.33) for $\epsilon = 0$ and, by the property of differentiability with respect to parameters, there exists a bounded $L_1(k,\epsilon)$ such that

$$L(k,\epsilon) = L_0(k) + \epsilon L_1(k,\epsilon) \tag{3.37}$$

The transformation (3.32) decomposes the system (3.31) into a slow $\theta-$system

$$\begin{aligned}\theta(k+1) &= (I-\epsilon g(k,\epsilon)\gamma(k)[d\phi^T(k)+c^T L(k,\epsilon)])\theta(k) \\ &\quad + \epsilon g(k,\epsilon)\gamma(k)c^T z(k)\end{aligned} \tag{3.38}$$

and a fast $z-$system

$$z(k+1) = (A+\epsilon g(k,\epsilon)[L(k+1,\epsilon)-b\phi^T(k)]\gamma(k)c^T)z \tag{3.39}$$

For a stability analysis the forcing term in (3.38) can be ignored and the analysis can be performed on

$$\theta(k+1) = [I - \epsilon g(k,\epsilon)\gamma(k)v^T(k,\epsilon)]\theta(k) \tag{3.40}$$

where the signal

$$v^T(k,\epsilon) = d\phi^T(k) + c^T L(k,\epsilon) \tag{3.41}$$

can be approximated for sufficiently small ϵ by the signal

$$v_0^T(k) \;=\; d\phi^T(k) + c^T L_0(k) \tag{3.42}$$

already defined in (2.22). For small ϵ the stability properties of (3.40) will therefore be determined by the stability properties of the simplified system

$$\theta(k+1) \;=\; [I - \epsilon\gamma(k)v_0^T(k)]\theta(k) \tag{3.43}$$

where we approximated $g(k,\epsilon) = 1+O(\epsilon)$ (cf. 2.18). Notice that for sufficiently small ϵ the fast system is exponentially stable (because A is exponentially stable (cf. 3.29)). Also the original system (3.31) inherits the stability properties of (3.38) and (3.39) because the transformation (3.32) preserves the stability properties. Hence for sufficiently small ϵ the system (3.31) has the same stability as the approximate slow system (3.43). The justification of this statement is the topic of the next sections.

3.3.4 Existence of L (Discrete Time)

To prove and interpret the state space decomposition result for the discrete-time system (3.31), we now prove the existence of $L(k,\epsilon)$.

Lemma 3.4:

Assuming that the $m \times m$ matrix A is Hurwitz (condition (3.29)) and that the regressor $\phi(k)$ is bounded with norm Φ (condition (2.12)) as well as the gain sequence $\gamma(k)$ with norm Γ (condition (2.15)), there exist constants ϵ^* and l^* such that for all $\epsilon \in (-\epsilon^*,\epsilon^*)$ there exists a unique $m \times n$ matrix $L(k,\epsilon)$ solution to

$$\begin{aligned} L(k+1,\epsilon) \;&=\; AL(k,\epsilon)+b\phi^T(k)+\epsilon[L(k+1,\epsilon)-b\phi^T(k)] \\ &\quad \cdot g(k,\epsilon)\gamma(k)[d\phi^T(k) + c^T L(k,\epsilon)] \end{aligned} \tag{3.33}$$

such that

$$L(k,\epsilon) = L_0(k) + \epsilon L_1(k,\epsilon) \tag{3.37}$$

where $L_0(k)$ satisfies

$$L_0(k) = AL_0(k) + b\phi(k) \tag{3.35}$$

Furthermore, $L(k,\epsilon)$ and $L_1(k,\epsilon)$ are bounded by

$$\|L(\cdot,\epsilon)\| \leq (1 + |\epsilon|l^*)\|L_0\| \ , \ \|L_1(\cdot,\epsilon)\| \leq l^*\|L_0\| \tag{3.44}$$

Using transformation (3.32) with $L(k,\epsilon)$ defined as above, the system (3.31) decomposes into a fast subsystem (3.39) and a slow subsystem (3.40) which completely characterize the stability properties of the original system (3.31).

Remark: Bounds ϵ^*, l^* are found by optimizing the range of allowable ϵ over the inequalities

$$\epsilon^2 l^2 \frac{\Gamma K}{1-a}|c|\|L_0\|^2 + |\epsilon|l\|L_0\|\frac{\Gamma K}{1-a}\left(2|c|\|L_0\| + (|b||c|+|d|)\Phi + d\Phi\frac{1-a}{K}\right)$$

$$+ (\|L_0\| + |b|\Phi)(|c|\|L_0\| + |d|\Phi)\frac{\Gamma K}{1-a} - l\|L_0\| \leq 0 \tag{3.45}$$

$$2\epsilon^2 l \frac{\Gamma K}{1-a}|c|\|L_0\| + |\epsilon|\frac{\Gamma K}{1-a}\left(2|c|\|L_0\|\right.$$

$$\left. + (|b||c|+|d|)\Phi + |d|\Phi\frac{1-a}{K}\right) - 1 \leq 0 \tag{3.46}$$

and

$$|\epsilon||d|\Phi\Gamma < 1 \tag{3.30}$$

As the stability analysis will further constrain the possible range of ϵ, we postpone this optimization until Section 3.5, where we discuss the stability properties in detail.

Proof.

In the Banach space $L(\mathbf{Z})$ of bounded sequences of $m \times n$ matrices equipped with the supremum norm consider the closed subsets $L_l(\mathbf{Z})$ of uniformly bounded sequences

$$L_l(\mathbf{Z}) = \{Y \in L(\mathbf{Z}) : \|Y\| \leq (1+|\epsilon|l)\|L_0\|\},\ l > 0 \tag{3.47}$$

and define on $L_l(\mathbf{Z})$ the operator W:

$$W(Y)(k) = \epsilon \sum_{j=-\infty}^{k} A^{k-j} G(Y(j), Y(j+1), j, \epsilon) + L_0(k) \tag{3.48}$$

where

$$G(Y_1, Y_2, k, \epsilon) = (Y_2 - b\phi^T(k)) g(k,\epsilon) \gamma(k) (\phi^T(k) d + c^T Y_1) \tag{3.49}$$

Noting that for all $Y_1, Y_2 \in L_l(\mathbf{Z})$

$$|\epsilon G(Y_1, Y_2, k, \epsilon)| \leq \frac{|\epsilon| \Gamma \Phi}{1 - |\epsilon||d| \Gamma \Phi} ((1 + |\epsilon| l) \|L_0\| + |b| \Phi)((1 + |\epsilon| l) |c| \|L_0\| + |d| \Phi) \tag{3.50}$$

and

$$|\epsilon G(Y_1(k), Y_1(k+1), k, \epsilon) - \epsilon G(Y_2(k), Y_2(k+1), k, \epsilon)|$$

$$\leq \frac{|\epsilon| \Gamma \Phi}{1 - |\epsilon||d| \Gamma \Phi} (2(1 + |\epsilon| l) |c| \|L_0\| + (|d| + |b||c|) \Phi) \|Y_1 - Y_2\| \tag{3.51}$$

we conclude from (3.50) that W is well defined for all ϵ and l satisfying (3.45), while from (3.51) we see that, for all ϵ and l satisfying (3.46),

$$\|W(Y_1) - W(Y_2)\| < \|Y_1 - Y_2\| \tag{3.52}$$

Hence, W is a contraction operator with the (unique) fixed point $L(k,\epsilon)$, which satisfies (3.33) and (3.43) by construction.

□

Remark: Since the condition (3.46) implies (3.30), the latter need not be separately assumed. In many applications the sign of d is known and equals the sign of ϵ and also $\phi^T(k)\gamma(k) \geq 0$ for all k, especially when $\gamma(k) \equiv \phi(k)$. In such applications the inverse (2.17) can always be explicitly evaluated and the explicit form (3.31) always exists. Then, starting from (3.48), less conservative bounds can be obtained taking into account that $|g(k,\epsilon)| \leq 1$.

The analogy with the continuous-time case is complete and, as in the remark after Lemma 3.3, we note that (3.45) implies the exponential stability of the $z-$system (3.39).

3.3.5 Comparison of the Two Decompositions

We conclude this discussion of the time-scale decomposition by comparing the convolution operator decomposition of Lemma 3.1 with the state-space decomposition of Lemma 3.3. A state representation of the perturbation operator Δ of Lemma 3.1,

$$\Delta(\theta)(t) = \phi(t)V(\theta)(t) - \phi(t)v_0^T(t)\theta(t) \tag{3.53}$$

shown in Figure 3.1b, is obtained if $L_0(t)$ is used instead of $L(t,\epsilon)$ in the transformation (3.6). Thus substituting

$$z_0(t) = x(t) - L_0(t)\theta(t) \tag{3.54}$$

into (3.4) and, noting from (3.11) that

$$\dot{L}_0(t) = AL_0(t) + b\phi^T(t) \tag{3.55}$$

we obtain

$$\dot{\theta}(t) = -\epsilon\phi(t)v_0^T(t)\theta(t) - \epsilon\phi(t)c^T z_0(t) \tag{3.56}$$

$$\dot{z}_0(t) = (A+\epsilon L_0(t)\phi(t)c^T)z_0(t) + \epsilon L_0(t)\phi(t)v_0^T(t)\theta(t) \tag{3.57}$$

which shows that a realization of the operator Δ is the triple

$$\begin{aligned} A_\Delta &= A+\epsilon L_0(t)C_\Delta(t) \\ B_\Delta &= \epsilon L_0(t)\phi(t)v_0^T(t) \\ C_\Delta &= \phi(t)c^T \end{aligned} \tag{3.58}$$

This representation of the system in Figure 3.1b is given in Figure 3.3. The fact that $|B_\Delta|$ is $O(\epsilon)$ shows that the gain of Δ is $O(\epsilon)$, as established by (2.6) in Lemma 3.1.

In what follows, the stability analysis is performed for a finite dimensional representation of the transfer function H. However, the

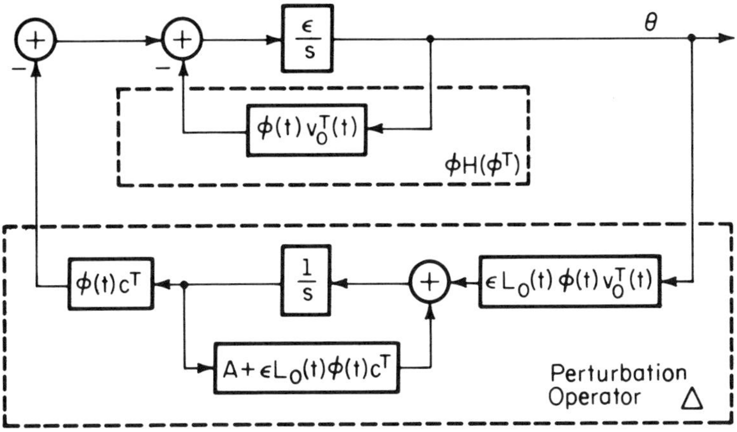

Fig. 3.3 A state-space representation of the perturbation operator Δ.

simple relationship between the operator Δ and the transformation L indicates that the same analysis remains valid for infinite dimensional systems in which Δ is a well defined $\epsilon-$perturbation.

3.4 AVERAGING ANALYSIS

The time-scale decomposition has demonstrated that the dominant part of the slow adaptation loop in Figure 3.1a is the inner loop in Figure 3.1b described by

$$\dot{\theta} = -\epsilon R(t)\theta \tag{4.1}$$

in the continuous-time case, and by

$$\theta(k+1) = [I - \epsilon R(k)]\theta(k) \tag{4.2}$$

in the discrete-time case, where, in view of (2.8) and (3.15),

$$R(t) = \phi(t)(H(\phi^T))(t) = \phi(t)v_0^T(t) \tag{4.3}$$

and, in view of (2.21) and (3.43),

$$R(k) = \gamma(k)(H(\phi^T))(k) = \gamma(k)v_0^T(k) \tag{4.4}$$

We note that in some discrete-time algorithms $\gamma(k) = \phi(k)$ and the analogy of (4.3) and (4.4) is complete. This analogy remains helpful when the dependence of $\gamma(k)$ on $\phi(k)$ is more complicated, such as in the least square algorithms.

3.4.1 Sample Averages

To analyze the stability properties of linear time-varying systems (4.1) and (4.2), we first observe that, although ϕ and v_0, and hence R, vary rapidly, the variations of θ are slow, provided that ϵ is sufficiently small. This fact will allow us to formulate stability conditions for (4.1) and (4.2) in terms of *sample averages*

$$\bar{R}_l = \frac{1}{T_l}\int_{t_{l-1}}^{t_l} R(t)dt \quad , \quad t_l = t_{l-1} + T_l \tag{4.5}$$

in the continuous-time case, and

$$\bar{R}_l = \frac{1}{K_l}\sum_{k_{l-1}}^{k_l-1} R(k) \quad , \quad k_l = k_{l-1}+K_l \tag{4.6}$$

in the discrete-time case, over finite sample intervals

$$0 < T_m \leq T_l \leq T_M < \infty$$

$$0 < K_m \leq K_l \leq K_M < \infty \quad , \quad \forall i \in \mathbf{Z} \tag{4.7}$$

Let the state transition matrices of (4.1) and (4.2) be, respectively,

$$F(t,\tau), F(\tau,\tau) = I \quad \text{and} \quad F(k,j), F(j,j) = I \tag{4.8}$$

and denote by $\bar{F}_l$ both

$$\bar{F}_l := F(t_l,t_{l-1}) \quad \text{and} \quad \bar{F}_l := F(k_l,k_{l-1}) \tag{4.9}$$

and by θ_l both

$$\theta_i := \theta(t_i) \quad \text{and} \quad \theta_i := \theta(k_i) \tag{4.10}$$

In this unified notation the transition from θ_j to θ_i for both systems (4.1) and (4.2) is given by

$$\theta_i = \bar{F}_i \bar{F}_{i-1} \cdots \bar{F}_{i-j} \, , \quad j < i \, ; \quad j,i \in \mathbf{Z}$$

and the following lemma is obvious.

Lemma 3.5: Uniform Contractions

If $\bar{F}_i$ is a sequence of uniform contractions for all i, that is if there exists a positive constant α such that

$$|\bar{F}_i| \leq e^{-\alpha} \quad , \quad \forall \, i \in \mathbf{Z} \quad , \tag{4.11}$$

then the system (4.1), (or (4.2)), is exponentially stable. Conversely, if the system (4.1), (or (4.2)), is exponentially stable, then there exists a sequence of sample intervals for which $\bar{F}_i$, $i \in \mathbf{Z}$, is a sequence of uniform contractions.

□

Our next step is to approximate $\bar{F}_i$ in terms of $\bar{R}_i$. For the continuous-time case this is accomplished as follows.

Lemma 3.6: Averaging (Continuous Time)

Assume that $R(t)$ is an $n \times n$ integrable matrix function of t and let

$$\rho_i = \frac{1}{T_i} \int_{t_{i-1}}^{t_i} |R(t)| dt \tag{4.12}$$

Then $\bar{F}_i$ is given by

$$\bar{F}_i = I - \epsilon T_i \bar{R}_i + (\epsilon T_i)^2 M_i \tag{4.13}$$

and for all ϵ satisfying

$$|\epsilon| T_i \rho_i \leq \frac{1}{2} \tag{4.14}$$

matrix M_i is bounded by $|M_i| \leq 2\rho_i^2$.

Proof

In the Banach space $C(t_{l-1},t_l)$ of continuous, bounded, $n \times n$ matrix functions Y of t, equipped with the norm

$$\|Y\| = \sup_{t_{l-1} \leq t \leq t_l} |Y(t)| \tag{4.15}$$

the state-transition matrix $F(t_l,t_{l-1})$, $t \in [t_{l-1},t_l]$, is the fixed point of the operator

$$U(Y)(t) = I - \int_{t_{l-1}}^{t} \epsilon R(\tau)Y(\tau)d\tau \tag{4.16}$$

which for all $Y_1,Y_2 \in C(t_{l-1},t_l)$ satisfies

$$\|U(Y_1)\| \leq 1 + |\epsilon| T_l \rho_l \|Y_1\| < \infty \tag{4.17}$$

and

$$\|U(Y_1)-U(Y_2)\| \leq |\epsilon| T_l \rho_l \|Y_1-Y_2\| \tag{4.18}$$

Hence, for all ϵ bounded by (4.14), the sequence generated by

$$Y_k(t) = U(Y_{k-1}(t)) \quad , \quad Y_0(t) = I \tag{4.19}$$

converges to $F(t,t_{l-1})$,

$$\lim_{k \to \infty} Y_k(t) = F(t,t_{l-1}) \tag{4.20}$$

and the first term of this series:

$$Y_1(t) = I - \epsilon \int_{t_{l-1}}^{t} R(\tau)d\tau \tag{4.21}$$

approximates $F(t,t_{l-1})$ with the error bounded by

$$\|F(t,t_{l-1})-Y_1(t)\| \leq \sum_{k=2}^{\infty} |\epsilon|(T_l\rho_l)^{k-1}\|Y_1(t)-Y_0(t)\|$$

$$\leq 2(|\epsilon| T_l \rho_l)^2 < \frac{1}{2} \tag{4.22}$$

For $t = t_l$ this fact proves (4.13) with $|M_l| \leq 2\rho_l^2$.

□

For completeness we restate the same result in the discrete-time form.

Lemma 3.7: Averaging (Discrete Time)

Assume that $R(k)$ is a sequence of bounded $n \times n$ matrices and let

$$\rho_i = \frac{1}{K_i} \sum_{k=k_{i-1}}^{k_i-1} |R(k)| \tag{4.23}$$

Then, as in Lemma 3.6, $\bar{F}_i$ is given by (4.13) and, for ϵ satisfying (4.14), $|M_i| \leq 2\rho_i^2$.

Proof

With the operator

$$U(Y)(k) = I - \sum_{j=k_{i-1}}^{k-1} \epsilon R(j)Y(j) \ , \ k = k_{i-1}, k_{i-1}+1, \ldots, k_i \tag{4.24}$$

$U(Y)(k_{i-1}) = I$, in place of (4.16), the proof is the same as that of Lemma 3.6.

□

3.4.2 Stability Condition for Signals with Sample Averages

We are now in the position to formulate a sufficient condition for exponential stability applicable to both the continuous-time system (4.1) and the discrete-time system (4.2).

Theorem 3.1: Stability via Sample Averages

Consider (4.1) with $R(t)$ an $n \times n$ integrable matrix function of t (or (4.2) with $R(k)$ a sequence of bounded $n \times n$ matrices). If for a sequence of sample averages $\bar{R}_i$, defined for all $i \in \mathbf{Z}$ by (4.5), (or (4.6)), a constant positive definite matrix $P = P^T > 0$ can be found satisfying

$$P\bar{R}_i + \bar{R}_i^T P \geq I \ , \ \forall \ i \in \mathbf{Z} \tag{4.25}$$

then there exists an $\epsilon^* > 0$ such that the system (4.1), (or (4.2)), is exponentially stable for all $\epsilon \in (0,\epsilon^*)$.

Proof

To show that

$$\theta_i^T P \theta_i < \theta_{i-1}^T P \theta_{i-1} \quad , \quad \forall\, i \in \mathbf{Z}$$

we substitute $\bar{F}_i$ from (4.13) and obtain

$$\bar{F}_i^T P \bar{F}_i = P - \epsilon T_i (P\bar{R}_i + \bar{R}_i^T P - \epsilon T_i G_i)$$

where

$$G_i = PM_i + M_i^T P + \epsilon T_i (M_i^T P \bar{R}_i + \bar{R}_i^T P M_i) + (\epsilon T_i)^2 M_i^T P M_i$$

is bounded by $|G_i| \le 7\rho_i^2 |P| = p_i$. Hence for all $\epsilon T_i \in (0, p_i^{-1})$, the expression (4.26) implies that

$$\bar{F}_i^T P \bar{F}_i = P - \epsilon T_i (I - \epsilon T_i G_i) < P \tag{4.26}$$

Using Lemma 3.5 and $\epsilon^* = \inf_i p_i^{-1} > 0$ completes the proof.

□

Remark: It follows from the proof of Theorem 3.1 that the state transition matrix $F(t,0)$ of (4.1), satisfies

$$|F(t,0)| \le e^{\epsilon \rho_n T_n} \left(\frac{\lambda_{\max}(P)}{\lambda_{\min}(P)} \right)^{1/2} \prod_{k=1}^{n} a_k(\epsilon) \quad , \quad \forall\, t \in (t_{n-1}, t_n)$$

where $a_k(\epsilon)$ is evaluated from (4.26):

$$a_k(\epsilon) = \left(1 - \epsilon \frac{T_k}{\lambda_{\max}(P)} + (\epsilon T_k)^2 7 \frac{\rho_k^2 \lambda_{\max}(P)}{\lambda_{\min}(P)} \right)^{1/2}$$

Hence, we can overbound $F(t,\tau)$ as

$$|F(t,\tau)| \le K e^{-a(\epsilon)(t,\tau)} \quad , \quad \forall\, t \ge \tau$$

with $K = \sqrt{e(\lambda_{\max}(P)/\lambda_{\min}(P))}$, with $e = 2.71828\ldots$, because of (4.14) and with

$$a(\epsilon) \leq - \lim_k \sup \, (2\sum_{l=1}^{k} T_l)^{-1} \sum_{l=1}^{k} \ln a_k(\epsilon) \tag{4.27}$$

which, using $\rho_M = \sup_k \rho_k$, can be estimated as

$$a(\epsilon) \leq - \frac{1}{2T_M} \ln \left(1 - \epsilon \frac{T_m}{\lambda_{\max}(P)} + (\epsilon T_M)^2 7 \frac{\rho_M^2 \lambda_{\max}(P)}{\lambda_{\min}(P)} \right) \tag{4.28}$$

□

The sufficient stability condition of Theorem 3.1 is a general *signal dependent positivity condition,* which applies to large classes of system matrices $R(t)$ or $R(k)$ and, hence, to large classes of signals $\phi(t)$ or $\phi(k)$. For more restrictive classes of signals, much sharper stability conditions can be obtained.

3.4.3 Stability-Instability Boundary for Periodic Signals

Particularly important are the necessary and sufficient conditions for $T-$periodic or $K-$periodic‡ signals ϕ when

$$R(t+T) = R(t) \quad \text{or} \quad R(k+K) = R(k) \quad .$$

In this case all the sample intervals are the same, $T_l = T$ or $K_l = K$, and all the sample averages are equal, $R_l = R$, $\rho_l = \rho$, $\forall i \in \mathbf{Z}$. We will use the well known stability theorem for periodic systems, found, for example in Hale (1980) and Miller and Michel (1982).

Theorem 3.2:

Let $R(t)$ be an $n \times n$ matrix $T-$periodic integrable function of t, (or let $R(k)$ be a $K-$periodic sequence of bounded $n \times n$ matrices). Then the eigenvalues λ_i, $i = 1,\ldots,n$, of the state transition matrix $F(t+T,t)$ (or $F(k+K,k)$) are independent of t (of

‡ It should be clear from the context when K is used to denote periodicity and when it represents an exponential bound constant (as in the preceding Remark).

k) and the system (4.1) (or (4.2)) is *exponentially stable if and only if*

$$\max_i |\lambda_i(F(t+T,t)| < 1 \text{ , } \quad (\text{or } \max_i |\lambda_i(F(k+K,k)| < 1) \tag{4.29}$$

and is *unstable* if

$$\max_i |\lambda_i| > 1 \tag{4.30}$$

□

In combination with the approximation (4.13), this theorem yields the following *necessary and sufficient condition* for exponential stability.

Theorem 3.3: Stability

For periodic $R(t)$, (or $R(k)$), as in Theorem 3.2, there exists a positive constant ϵ^* such that the system (4.1) (or (4.2)) is *exponentially stable* for all $\epsilon \in (0,\epsilon^*)$ *if and only if*

$$\min_i \text{Re}\lambda_i(\bar{R}) > 0 \tag{4.31}$$

Proof

By Lemma 3.6 and in view of the continuous dependence of the eigenvalues of a matrix on its entries, we have

$$\lambda_i(F(t+T,t)) = 1 - \epsilon T[\lambda_i(\bar{R}) - \delta(\epsilon T)] \tag{4.32}$$

where $\delta(\epsilon T) \to 0$ as $\epsilon T \to 0$. Note, however, that if $\bar{R}$ is diagonalizable (4.32) can be improved by using $O(\epsilon T)$ instead of $\delta(\epsilon T)$. In terms of $\text{Re}\lambda_i(\bar{R})$ and $\text{Im}\lambda_i(\bar{R})$ expression (4.32) yields

$$\begin{aligned} |\lambda_i(F(t+T,t))|^2 &= [1-\epsilon T\text{Re}\lambda_i(\bar{R})+\epsilon T\delta(\epsilon T)]^2 \\ &+ [\epsilon T\text{Im}\lambda_i(\bar{R})+\epsilon T\delta(\epsilon T)]^2 \end{aligned} \tag{4.33}$$

and, hence,

$$|\lambda_i(F(t+T,t))| = 1-\epsilon T[\text{Re}\lambda_i(\bar{R})-\delta(\epsilon T)] \tag{4.34}$$

Therefore there exists $\epsilon^* > 0$ such that for all $\epsilon \in (0,\epsilon^*)$

$$|\lambda_i(F(t+T,t)| \; < \; 1 \text{ if and only if } \mathrm{Re}\lambda_i(\bar{R}) \; > \; 0$$

which, by Theorem 3.2, completes the proof.

□

Remark: An estimate of ϵ^* and the convergence rate $a(\epsilon)$ can be obtained by specializing Theorem 3.1 and (4.28) for the periodic case: $T_M = T_m = T$, $\rho_M = \rho$, and defining P as the unique solution of

$$\bar{R}^T P + P\bar{R} \; = \; I \tag{4.35}$$

The expression (4.26) then yields

$$\epsilon^* \; = \; \frac{1}{7\rho^2 T|P|} \tag{4.36}$$

We see that for all $\bar{R}$ which are well-conditioned in the sense that $\rho|P| \sim 1$, the product $\epsilon^* T\rho$ is constant. This constant is the intrinsic "gain coefficient" of the adaptive loop.

The expression (4.34) and Theorem 3.2 also prove the following sufficient condition for instability.

Theorem 3.4: Instability

For $R(t)$ (or $R(k)$), as in Theorem 3.2, if

$$\min_i \mathrm{Re}\lambda_i(\bar{R}) \; < \; 0 \tag{4.37}$$

then there exists an $\epsilon^* > 0$ such that for all $\epsilon \in (0,\epsilon^*)$ the periodic system (4.1) (or (4.2)), is unstable.

□

From the last two theorems we see that if $\mathrm{Re}\lambda_i(\bar{R}) \; \neq \; 0$, $\forall\, i = 1,\ldots,n$, then *the eigenvalues of* $\bar{R}$ *delineate a sharp stability-instability boundary for periodic systems* (4.1) and (4.2). This fact will be used to develop practical stability criteria for slow adaptation in Section 3.5, and to apply them to a couple of examples in Sections 3.6 and 3.7.

3.5 STABILITY CRITERIA FOR SLOW ADAPTATION

We are prepared to formulate stability and instability conditions for the slow adaptation system

$$\begin{bmatrix} \dot{\theta} \\ \dot{x} \end{bmatrix} = \begin{bmatrix} -\epsilon d\phi(t)\phi^T(t) & -\epsilon\phi(t)c^T \\ b\phi^T(t) & A \end{bmatrix} \begin{bmatrix} \theta \\ x \end{bmatrix} \tag{5.1}$$

and its discrete-time counterpart

$$\begin{bmatrix} \theta(k+1) \\ x(k+1) \end{bmatrix} = \tag{5.2}$$

$$\begin{bmatrix} I-\epsilon g(k,\epsilon)d\gamma(k)\phi^T(k) & -\epsilon g(k,\epsilon)\gamma(k)c^T \\ b\phi^T(k)g(k,\epsilon) & A-\epsilon g(k,\epsilon)\phi^T(k)d\gamma(k)bc^T \end{bmatrix} \begin{bmatrix} \theta(k) \\ x(k) \end{bmatrix}$$

To avoid unneccessary repetitions, the discussion will focus on the continuous-time system (5.1). Our three main tools for the stability analysis are: *first*, the decomposition of (5.1) into the fast $z-$system

$$\dot{z} = (A+\epsilon L(t,\epsilon)\phi^T(t)c^T)z \tag{5.3}$$

and the slow $\theta-$system

$$\dot{\theta} = -\epsilon\phi(t)v^T(t,\epsilon)\theta = -\epsilon[\phi(t)v_0^T(t)+\epsilon\phi(t)c^T L_1(t,\epsilon)]\theta \tag{5.4}$$

where the forcing term $-\epsilon\phi(t)c^T z$ has been deleted from (5.4) as inessential; *second*, the stability results for the unperturbed part of the $\theta-$system (5.4) obtained by averaging; and, *third*, the bounds on the perturbation terms imposed by the Total Stability Theorem 1.4 given in Chapter 1.

3.5.1 Signal Dependent Stability Criteria

In the general case our results concerning exponential stability and instability are stated as sufficient conditions. For the case of periodic signals, $\phi(t) = \phi(t+T)$, the exponential stability conditions are both necessary and sufficient. We also give a frequency domain interpretation of these conditions.

Theorem 3.5: Signal Dependent Stability

Let $\phi(t)$ be an $n-$vector integrable function of t (or $\phi(k)$ a bounded sequence of $n-$vectors). Assume that the unperturbed $z-$system and unperturbed $\theta-$system are exponentially stable: $\max_i \mathrm{Re}\lambda_i(A) < 0$ (or $\max_i |\lambda_i(A)| < 1$), $i=1,\ldots,m$ and the Sample Average Theorem (Theorem 3.1 with $R(t) = \phi(t)v_0^T(t)$) holds for a sequence of sample intervals T_j, $j \in \mathbf{Z}$.

Then there exists a positive constant ϵ^* such that the system (5.1) (or (5.2)) is exponentially stable for all $\epsilon \in (0,\epsilon^*)$.

□

Remark: The bound ϵ^* can be found by optimizing with respect to l, over the inequalities

$$\epsilon^2 l\alpha_1 + |\epsilon|\alpha_2 + \frac{\alpha_3}{l} - 1 \leq 0 \tag{5.5}$$

$$2\epsilon^2 l\alpha_1 + |\epsilon|\alpha_2 - 1 \leq 0 \tag{5.6}$$

$$\sqrt{e}\left(\frac{\lambda_{\max}(P)}{\lambda_{\min}(P)}\right)^{1/2} |c|\epsilon^2\|L_0\| l\Phi - a(\epsilon) \leq 0 \tag{5.7}$$

$$0 \leq \epsilon T_M \rho_M \leq \frac{1}{2} \tag{5.8}$$

where $\sqrt{e} = (2.71828\ldots)^{1/2}$ and $a(\epsilon)$ is given by (4.27) or (4.28). α_1,α_2 and α_3 are

$$\alpha_1 = |c|\|L_0\|\Phi\,\frac{K}{a}$$

$$\alpha_2 = 2\alpha_1 + |d|\Phi^2\,\frac{K}{a}$$

$$\alpha_3 = \alpha_1 + |d|\Phi^2\,\frac{K}{a}$$

Analogous expressions are valid for the discrete time case.

It is clear that for any given $l > \alpha_3$, a sufficiently small $\epsilon > 0$ exists which satisfies (5.5), (5.6), (5.7), and (5.8). It is also possible to obtain an explicit expression for the best bound ϵ^*. However, this expression is rather complicated. Instead, it is simpler to perform a numerical maximization of ϵ in every particular application.

Proof

As already pointed out in the remark after Lemma 3.3, the bound $\beta K < a$ of Theorem 1.2, Chapter 1, applied to the perturbed $z-$system, is less restrictive than (5.5) and, hence, (5.5) guarantees that the $z-$system is exponentially stable. To prove that the perturbed $\theta-$system is also exponentially stable, note that the bound (5.8) is constructed to satisfy both Theorem 3.1 for the unperturbed $\theta-$system and the bound $\beta K < a$ of Theorem 1.3 for the perturbed $\theta-$system. A simultaneous satisfaction of the inequalities (5.5) and (5.6), which by Lemma 3.3 (or 3.4) guarantees the validity of the decomposition and of the exponential stability bound (5.7) proves the theorem.

□

For instability of (5.1) (or (5.2)), it is sufficient that at least one signal $\phi(t)$ (or $\phi(k)$) exists which makes the unperturbed $\theta-$system, and, hence, the whole slow adaptation system, unstable for sufficiently small ϵ.

Theorem 3.6: Signal-Dependent Instability

If the transfer function

$$H(s) = d + c^T(sI-A)^{-1}b \tag{5.9}$$

(or its discrete-time counterpart $H(z)$) is not strictly positive real, then there exists $\epsilon^* > 0$ and a signal $\phi(t)$ (or $\phi(k)$) such that (5.1), (or (5.2)) is unstable for all $\epsilon \in (0,\epsilon^*)$.

Proof

By Theorem 3.3, it is sufficient to find a periodic signal ϕ such that

$$\min_i \mathrm{Re}\lambda_i(\bar{R}) < 0 \quad , \quad Re\lambda_i(\bar{R}) \neq 0 \ , \ i = 1,\ldots,n \qquad (5.10)$$

where, from (4.3)-(4.6),

$$\bar{R} = \frac{1}{T}\int_t^{t+T} \phi(\tau)v_0^T(\tau)d\tau \ , \ \text{or} \ \bar{R} = \frac{1}{K}\sum_{l=k}^{k+K-1} \gamma(l)v_0^T(l)$$

If we let each component ϕ_i of the $n-$vector ϕ be a periodic function of period T/i (or K/i, where K is the least common multiple of the first n integers) for all $i = 1,\ldots,n$, then $\bar{R}$ is a diagonal matrix, because the average over T of each product $\phi_i v_{oj}$ is zero if $i \neq j$. When H is not SPR, then the period T can be chosen such that for some i, say $i = l$, the average of $\phi_l v_{ol}$ over T is negative, while the averages of $\phi_i v_{ol}$ for $i \neq l$ are all nonzero. Such a signal ϕ satisfies (5.10).

□

3.5.2 Stability Criterion for Periodic Signals

We next give our main exponential stability result for the case of periodic signals.

Theorem 3.7:

Let $\phi(t)$ be an $n-$vector $T-$periodic integrable funtion of t (or $\phi(k)$ be a $K-$periodic sequence of bounded $n-$vectors) and assume that $\max_i \mathrm{Re}\lambda_i(A) < 0$ (of $\max_i|\lambda_i(A)| < 1$). Then provided that $\mathrm{Re}\lambda_i(\bar{R}) \neq 0$, $i = 1,\ldots,n$, there exists a positive constant ϵ^* such that the system (5.1) (or (5.2)) is exponentially stable for $\epsilon \in (0,\epsilon^*)$ if and only if

$$\min_i \mathrm{Re}\lambda_i(\bar{R}) > 0 \ , \quad i = 1,\ldots,n \qquad (5.11)$$

Remark: Bound ϵ^* can be obtained by optimizing the range of ϵ's with respect to l over the inequalities

$$\epsilon^2 l\alpha_1 + |\epsilon|\alpha_2 + \frac{\alpha_3}{l} - 1 \leq 0 \qquad (5.12)$$

$$\epsilon^2 2l\alpha_1 + |\epsilon|a_2 - 1 \leq 0 \qquad (5.13)$$

$$\epsilon^2 \left(e \, \frac{\lambda_{max}(P)}{\lambda_{min}(P)} \right)^{1/2} |c| \|L_0\| l\phi - |a(\epsilon)| \leq 0 \qquad (5.14)$$

$$0 < \epsilon T\rho < \frac{1}{2} \qquad (5.15)$$

where $e = 2.71828...$ and P is the solution of (4.35) and

$$\alpha_0 = 7\rho^2 \frac{\lambda_{max}(P)}{\lambda_{min}(P)}$$

$$\alpha_1 = |c| \|L_0\| \Phi \frac{K}{a}$$

$$\alpha_2 = 2\alpha_1 + |d|\Phi^2 \frac{K}{a}$$

$$\alpha_3 = \alpha_1 + |d|\Phi^2 \frac{K}{a}$$

$$\alpha(\epsilon) = -\frac{1}{2T} \ln \left[1 - \frac{\epsilon T}{\lambda_{max}(P)} + (\epsilon T)^2 \alpha_0 \right]$$

Completely analogous expressions are valid for the discrete-time case.

Proof

First notice that it is possible to satisfy the inequalities (5.12)-(5.15) for ϵ positive and sufficiently small and $l > \alpha_3$. Inequalities (5.12) and (5.13) guarantee that the decomposition into the $z-$system and the $\theta-$system is valid and, as in Theorem 3.5, that the perturbed $z-$system is exponentially stable. The inequality (5.14) simultaneously satisfies the exponential stability bound $\beta K < a$ of Theorem 3.3 for the unperturbed $\theta-$system and the bound on the perturbation imposed by the Total Stability Theorem (Theorem 1.3 in Chapter 1). This proves the sufficiency part of the Theorem. The necessity follows from the fact that if (5.11) is violated by $\mathrm{Re}\lambda_j(\bar{R}) < 0$ for some j, Theorem 3.4 implies that the unperturbed θ-system is unstable, which means that the perturbed system is unstable for $\epsilon > 0$ sufficiently small.

□

As in Theorem 3.5, the inequalities (5.12), (5.13) and (5.14) can be used to find the least conservative bound on ϵ for which the exponential stability can still be guaranteed using the condition (5.11). On the other hand, condition (5.11) is equivalent to the existence of a positive definite matrix $P = P^T$ such that

$$P\bar{R} + \bar{R}^T P > 0 \tag{5.16}$$

Requiring that (5.16) be satisfied with some special choices of P we can construct easily interpretable sufficient conditions for (5.16) to hold.

3.5.3 Average SPR Condition

A condition, which is extremely simple to interpret, is obtained when (5.16) is satisfied by $P = I$, namely

$$\bar{R} + \bar{R}^T > 0 \tag{5.17}$$

To show the meaning of this condition in terms of the frequency content of the signal ϕ and the properties of the transfer function H, we let

$$\phi(t) = \sum_{i=-\infty}^{\infty} \phi_i e^{j\omega_i t} \quad , \quad \omega_i = \frac{2\pi i}{T} \tag{5.18}$$

where the n-vector ϕ_{-i} is the conjugate $\bar{\phi}_i$ of the complex vector ϕ_i, that is,

$$\phi_{-i} = \bar{\phi}_i \quad , \quad \forall\, i \in \mathbf{Z} \tag{5.19}$$

When $\phi(t)$ is the input into H, the output is

$$v_0(t) = \sum_{i=-\infty}^{\infty} \phi_i H(j\omega_i) e^{j\omega_i t} \tag{5.20}$$

and the average (5.10) for $t \in [0,T]$ is

$$\bar{R} = \frac{1}{T}\int_0^T \phi(t)v_0^T(t)dt = \sum_{l=-\infty}^{\infty} H(j\omega_l)\phi_l\bar{\phi}_l^T \tag{5.21}$$

We note that, in general, $\bar{R}$ is not symmetric because the components of $\phi(t)$ with the same ω_l may have a phase shift. This observation is important in applications when some components of $\phi(t)$ appear as filtered versions for the same signal, as will be illustrated by examples in the next two sections.

For $\bar{R}$ of (5.21) to be nonsingular it is necessary that

$$\sum_{l=-\infty}^{\infty} \phi_l\bar{\phi}_l^T > 0 \tag{5.22}$$

which shows that $\phi(t)$ must be persistently exciting. Applying the condition (5.17) to an $\bar{R}$ defined by (5.21) we obtain a *signal dependent "average SPR" condition:*

$$\sum_{l=-\infty}^{\infty} \text{Re}H(j\omega_l)\text{Re}\phi_l\bar{\phi}_l^T > 0 \tag{5.23}$$

Clearly, (5.23) is satisfied if $\phi(t)$ is persistently exciting and if H is SPR. However, (5.23) is much less restrictive than the SPR condition. Condition (5.23) allows $\text{Re}H(j\omega_l) < 0$ for some ω_l, provided that in the sum of (5.23) the effect of the terms with $\text{Re}H(j\omega_l) > 0$ prevails. Given a non-SPR transfer function H, the condition (5.23) imposes a restriction on the amount of the "*ReH*-weighted energy" of ϕ in the range of frequencies where $\text{Re}H < 0$ and requires a larger amount of this "energy" in the range where $\text{Re}H > 0$. For signals which have desirable sprectral properties a non-SPR transfer function H behaves, *on average,* as an SPR transfer function. We have thus formulated a simple, albeit conservative, sufficient condition for exponential stability of (5.1).

Theorem 3.8:

If $\phi(t)$ is an n-vector T-periodic integrable function of t, and if the transfer function H is stable and satisfies the *average SPR condition* (5.23), then there exists $\epsilon^* > 0$ such that the slow adaptation system (5.1) is *exponentially stable* for all $\epsilon \in (0,\epsilon^*)$. (A verbatim analogue is valid for the discrete-time case.)

□

This condition can be used as a criterion to select a signal ϕ if some properties of a non-SPR transfer function H are known a priori. Typically $\mathrm{Re}H(j\omega)$ is negative at higher frequencies, say $\omega > \omega_c$. The criterion then requires that ϕ be persistently exciting in the range of lower frequencies $\omega < \omega_c$. This kind of a priori information is also useful in the design of low-pass filters to improve the stability properties of slow adaptation.

3.6 A LINEAR EXAMPLE AND DISCUSSION

Although most adaptive systems are nonlinear, in this section we avoid the issue of linearization and its validity by considering the example of an extremely simple continuous-time adaptive system which, in its exact form, is linear. This example will serve as an introduction to Section 3.7 and Chapters 4, 5, and 6, which analyze specific adaptive algorithms.

3.6.1 A Non-Minimum-Phase Plant

Consider the output error adaptive system in Figure 3.4 with only one adjustable parameter q. The adaptation law is

$$\dot{q} = -\epsilon r(t)e(t) \tag{6.1}$$

where $e(t)$ is the output error

$$e(t) = y(t) - \hat{y}(t) \tag{6.2}$$

and $r(t)$ is a reference input signal. This system is representative of

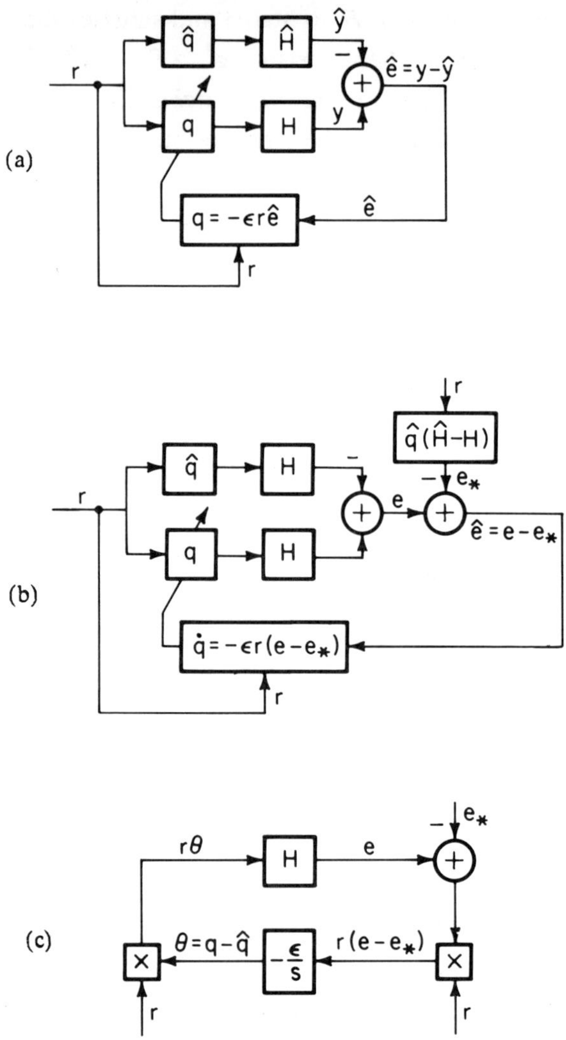

Fig. 3.4 The output error adaptive system.

two classes of adaptation problems. First, if $\hat{q}$ and the transfer function $\hat{H}$ represent an unknown plant, then the scheme in Figure 3.4a is an identification algorithm whereby q is being adjusted in order to identify the unknown $\hat{q}$. Whether the exact identification will be achieved or not depends on the transfer functions H and $\hat{H}$.

The second application of the same scheme is when q and H represent a plant to be controlled and $\hat{q}$ and $\hat{H}$ form a reference model to be matched. Again, the matching may or may not be achieved depending on H and $\hat{H}$.

In both applications the stability properties of this scheme are determined only by the feedback loop involving H and the adaptation law (6.1). In Figures 3.4b and 3.4c we add and subtract H to form the "tuned error" e_* which is caused only by the mismatch $\hat{H}-H$. We also introduce the parameter error $\theta = q-\hat{q}$. Clearly, if the tuned error e_* is zero, $\hat{H} = H$, and if the loop in 3.4c is exponential stable, then $\theta(t) \to 0$, $e(t) \to 0$ as $t \to \infty$. Under a rather unrealistic assumption that $\hat{H}$ is known and H is at our disposal, we can achieve $e_* = 0$ by selecting $H = \hat{H}$. Should we do so if $\hat{H}$ is not SPR? The answer of the theory in the preceding section is "Yes, if $r(t)$ is also at our disposal." By selecting $r(t)$ and ϵ to satisfy the exponential stability conditions, we can achieve the convergence $\theta(t) \to 0$, that is, the exact parameter identification $q(t) \to \hat{q}$. If, however, $\hat{H}$ is unknown, a way to quarantee the exponential stability is to select H to be SPR. An obvious victim then is the identification accuracy, because a nonzero input $e_*(t)$ will not allow $q(t)$ to converge to $\hat{q}$.

While in this linear example the magnitude of e_* does not influence the stability properties of the system, this is not so in more realistic nonlinear problems, for which the total stability theorem imposes a severe constraint on the size of e_*. This warning is issued here to prevent misinterpretations or hasty generalizations of the linear analysis which follows. The more complex issues associated with nonlinear averaging and linearization are pursued in Chapter 4. For this analysis we let

$$\hat{H}(s) = H(s) \ , \ e_*(t) = 0 \ , \ \forall \ t \tag{6.3}$$

and assume that $H(s)$ is stable and finite-dimensional. Denoting $r(t) = \phi(t)$ and adopting a state representation of $H(s)$ we then formulate the slow adaptation system (5.1). To make the analysis both simple and meaningful, we let $H(s)$ be a first order transfer function with a right-half-plane zero,

$$H(s) = \frac{1-\mu s}{1+s} = -\mu + \frac{1+\mu}{1+s} \tag{6.4}$$

which is obviously non-SPR for all $\mu > 0$. We have thus constructed what seems to be the simplest adaptive system capable of demonstrating most of the complex stability-instability phenomena occurring in more realistic systems. A minimal realization of $H(s)$ in (6.4) is

$$\begin{array}{l} A = -1 \ , \ b = 1 + \mu \\ c = 1 \ , \ d = -\mu \end{array} \tag{6.5}$$

and, hence, the slow adaptation system (5.1) is

$$\begin{bmatrix} \dot{\theta} \\ \dot{x} \end{bmatrix} = \begin{bmatrix} \epsilon\mu\phi^2(t) & -\epsilon\phi(t) \\ (1+\mu)\phi(t) & -1 \end{bmatrix} \begin{bmatrix} \theta \\ x \end{bmatrix} \tag{6.6}$$

3.6.2 High Gain Instability

Before we apply the results of the preceding section, let us establish that for $\epsilon > 0$ sufficiently large this system can be unstable. When $\phi(t) = \phi$ constant, this is clear from the characteristic equation

$$\lambda^2 + (1-\epsilon\mu\phi^2)\lambda + \epsilon\phi^2 = 0 \tag{6.7}$$

which shows that the system is unstable for

$$\epsilon\phi^2 > \frac{1}{\mu} \tag{6.8}$$

For a general $\phi(t)$ we recall a theorem stating that $\dot{v} = M(t)v$ is unstable if $\int_0^T$ trace $M(t)dt \to \infty$ as $T \to \infty$. This shows that (6.6) is unstable if

$$\epsilon \limsup_{T \to \infty} \frac{1}{T} \int_0^T \phi^2(t)dt > \frac{1}{\mu} \tag{6.9}$$

We see that the instability occurs when the loop gain is larger than the right-half-plane zero $\frac{1}{\mu}$.

3.6.3 Stability-Instability Boundary

Restricting both $\mu > 0$ and $\epsilon > 0$ to be small we will not allow our system (6.6) to experience this type of high gain instability. Instead we analyze the "slow drift" of θ for ϵ small.

We let $\phi(t)$ be periodic, say $\phi(t) = \cos\omega t$. Then (6.6) becomes

$$\begin{bmatrix} \dot{\theta} \\ \dot{x} \end{bmatrix} = \begin{bmatrix} \epsilon\mu\cos^2\omega t & -\epsilon\cos\omega t \\ (1+\mu)\cos\omega t & -1 \end{bmatrix} \begin{bmatrix} \theta \\ x \end{bmatrix} \tag{6.10}$$

Using the decomposition method of Section 3.3, we obtain for thc slow $\theta-$system

$$\dot{\theta} = \epsilon(\mu\cos^2\omega t - L(t,\epsilon)\cos\omega t)\theta \tag{6.11}$$

and for the fast $z-$system

$$\dot{z} = (-1 + \epsilon L(t,\epsilon)\cos\omega t)z \tag{6.12}$$

L being the solution of

$$\dot{L} = -L+(1+\mu)\cos\omega t-\epsilon\mu L\cos^2\omega t+\epsilon L^2\cos\omega t \tag{6.13}$$

which for $\epsilon = 0$ reduces to

$$L(t,0) = \frac{1+\mu}{1+\omega^2}\cos\omega t + \frac{(1+\mu)\omega}{1+\omega^2}\sin\omega t \tag{6.14}$$

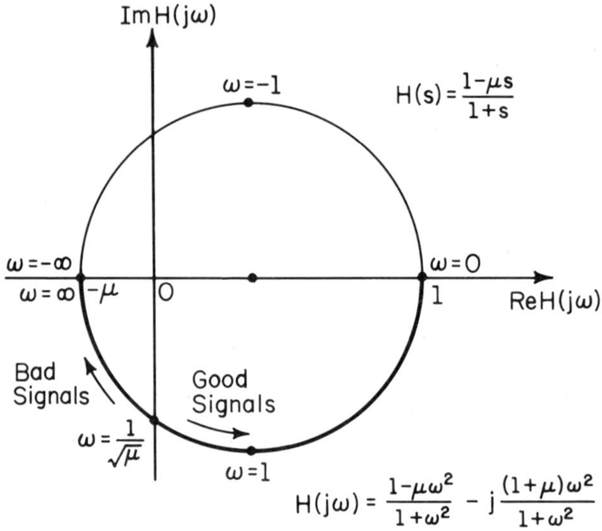

Fig. 3.5 The non-SPR transfer function *H(s)* with the single frequency stability boundary at $\omega = 1/\sqrt{\mu}$

In this example $\bar{R}$ is a scalar and the eigenvalue condition of Theorem 3.7 applies to $\bar{R}$ itself, namely to

$$\bar{R} = -\frac{\omega}{2\pi}\int_0^{2\pi/\omega}(\mu\cos^2\omega t - \cos\omega t L_0(t))dt \tag{6.15}$$

which gives

$$\bar{R} = \frac{1}{2}\frac{1-\mu\omega^2}{1+\omega^2} \tag{6.16}$$

Hence, (6.11) will be exponentially stable if and only if

$$\omega < \frac{1}{\sqrt{\mu}} \tag{6.17}$$

A rather far reaching interpretation of this condition is seen from Figure 3.5 where $\omega = 1/\sqrt{\mu}$ is the frequency at which the Nyquist plot of $H(j\omega)$ crosses the $j\omega$-axis, that is, Re$H(j\omega)$ becomes negative. Thus, (6.17) defines the single-frequency stability-instability boundary.

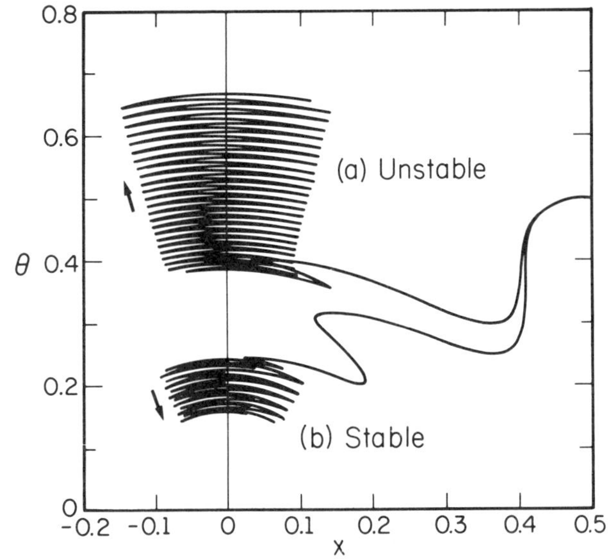

Fig. 3.6 Stability-instability boundary of the adaptive system (6.6).

If, instead of $\phi(t) = \cos\omega t$, a more general signal

$$\phi(t) = \sum_{l=1}^{N} r_l \sin(\omega_l t + \sigma_l) \tag{6.18}$$

is used, then the condition analogous to (6.17) is

$$\sum_{l=1}^{N} \frac{r_l^2}{1+\omega_l^2} > \sum_{l=1}^{N} \frac{r_l^2}{1+\mu^{-1}} \tag{6.19}$$

which is satisfied if every ω_l satisfies (6.17). However, (6.19) can also be satisfied if (6.17) is violated for some ω_l, provided that as a sum (6.19) still holds. This illustrates the statement that, in an average sense, $H(s)$ is required to behave as an SPR function.

We show in Figure 3.6 two simulated responses of the adaptive system with $\mu = 0.1$. As predicted by the criterion (6.17), the response (a) with $\phi_a(t) = \sin 4t$ is unstable because $\omega = 4 > 1/\sqrt{\mu} = \sqrt{10}$. According to (6.19), an additional low frequency component, namely the use of $\phi_b(t) = \sin 4t + 0.35 \sin t$, should lead to an exponentially stable response. Response (b) verifies this prediction and shows that the stability-instablility boundary delimited by our criterion is sharp.

3.6.4 Conservativeness of the Averaging Bound

Let us calculate the conditions of Theorem 3.7 in the case when $\phi(t) = \cos \omega t$. Under the condition (6.17), the adaptive system (6.10) is exponentially stable, if ϵ satisfies the system of inequalities

$$\epsilon^2 l \alpha_1 + |\epsilon|(2\alpha_1 + \alpha_2) + \frac{\alpha_1 + \alpha_2}{l} - 1 \leq 0 \tag{6.20}$$

$$\epsilon^2 2 l \alpha_1 + |\epsilon|(2\alpha_1 + \alpha_2) - 1 \leq 0 \tag{6.21}$$

$$\epsilon^2 l \sqrt{e} \, \frac{\alpha_1}{1+\mu} + \frac{\omega}{4\pi} \ln(1 - \epsilon\alpha_3 + \epsilon^2 \alpha_5) < 0 \tag{6.22}$$

$$0 < \epsilon \left(\frac{\alpha_3}{2} + \alpha_4 \right) - \frac{1}{2} < 0 \tag{6.23}$$

where

$$\alpha_1 = (1+\mu) l \, \frac{1+\omega}{1+\omega^2} \quad , \quad \alpha_2 = \mu(1+\mu)$$

$$\alpha_3 = \frac{2\pi}{\omega} \, \frac{1-\mu\omega^2}{2+\omega^2} \quad , \quad \alpha_4 = \frac{2(1+\mu)}{1+\omega^2} \tag{6.24}$$

$$\alpha_5 = (\alpha_3 + \alpha_4)(2\alpha_3 + 3\alpha_4) \quad , \quad e = 2.7128\ldots$$

These inequalities guarantee that the exact adaptive system inherits the stability properties of the simplified averaged equation

$$\dot{\theta}_{avg} = -\epsilon \bar{R} \theta_{avg} \tag{6.25}$$

and for this reason can be expected to be quite conservative. We first note that for $l > \alpha_1 + \alpha_2$ these inequatlities are obviously satisfied by a choice of a sufficiently small $\epsilon > 0$. The inequalities are particularly conservative as $\omega \to 0$, because then (6.22) and (6.23) require that $\epsilon \sim O(\omega)$. The result (6.8) for $\phi(t) = \phi = \text{constant}$ shows that, as the exponential stability is concerned, $\epsilon\phi^2$ can be almost as large as $1/\mu$. However, the averaging method guarantees not only this stability property, but also the $O(\epsilon)-$closeness of the true solution $\theta(t)$ to the average solution $\theta_{avg}(t)$. To show that for $\omega \to 0$ this requirement is restrictive, let us digress and examine the simple example:

$$\dot{\theta} = -\epsilon(\cos^2 \omega t)\theta \to \theta(t) = e^{-\epsilon t/2} \left(\epsilon^{-\frac{\epsilon}{4\omega}\sin 2\omega t} \right) \theta(0) \tag{6.26}$$

whose average $\theta_{avg}(t) = e^{-\epsilon t/2}\theta(0)$ correctly represents the exponential convergence of $\theta(t)$, but for $\omega << \epsilon$ fails to approximate $\theta(t)$.

Returning to the actual adaptive system (6.10), we give in Figure 3.7 two estimates of its stability region in the $(\omega,\epsilon)-$plane for $\mu = 0.1$. The A-region predicted by the averaging analysis is obtained as the largest area satisfying the inequalities (6.20)-(6.24). The $L-$region is obtained by a direct analysis of the $\theta-$system using the bounds on $L(t,\epsilon)$ estimated from (6.13) via (6.14). Clearly, the A-region is more conservative at lower frequencies, as explained above.

3.7 ANALYSIS OF AN OUTPUT ERROR ALGORITHM

Output error identification algorithms for discrete time will be discussed in greater detail in Section 5.1 of Chapter 5. Our aim here

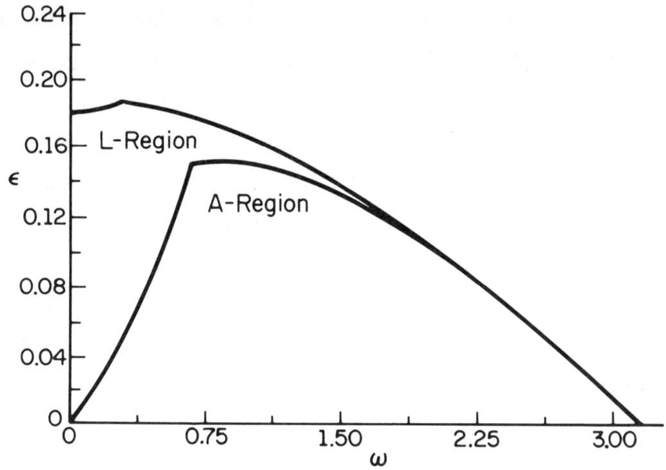

Fig. 3.7 Two estimates of the stability region: A - by averaging, L - by the time-scale decomposition.

is to illustrate some of the ideas of this chapter; to this end, we recall some of the basic issues of output error identification without any more discussion than necessary for the illustration. Output error algorithms are usually studied assuming that a relevant transfer function H is SPR, Landau (1979), Goodwin and Sin (1984).

In this section we will examine what happens to an identification algorithm when this transfer function is not SPR. The algorithm to be analyzed is an "A-class" identifier of Landau (1979), p. 191, also appearing in Goodwin and Sin (1984), pp. 83-87. For simplicity we let the proportional gain of Landau be zero and the integral gain be ϵI. Consistent with our interest in non-SPR transfer functions we delete the fixed moving-average filter (that is, we let $D = 1$ in Goodwin-Sin notation), because the goal of this filter is to guarantee that H is SPR. As it is not clear how to achieve this goal, our use of $D = 1$ is not a particularly unfair choice.

3.7.1 Output Error Algorithm

The update law is

$$\hat{\theta}(k+1) = \hat{\theta}(k) - \frac{\epsilon\bar{\phi}(k)}{1+\epsilon\bar{\phi}^T(k)\bar{\phi}(k)} (\bar{\phi}^T(k)\hat{\theta}(k)-y(k+1)) \quad (7.1)$$

where $y(k)$ is the output of the plant to be identified. The identifier output $\bar{y}(k)$, adjustable parameter vector $\hat{\theta}(k)$, and regressor vector $\bar{\phi}(k)$ are

$$\bar{y}(k) = \bar{\phi}^T(k-1)\hat{\theta}(k) \quad (7.2)$$

$$\hat{\theta}(k) = [-\hat{a}_1(k) \ -\hat{a}_2(k) \ \cdots \ -\hat{a}_n(k) \ \hat{b}_1(k) \ \cdots \ \hat{b}_r(k)]^T \quad (7.3)$$

$$\bar{\phi}(k) = [\bar{y}(k) \ \bar{y}(k-1) \ \cdots \ \bar{y}(k+1-n) \ u(k) \ \cdots \ u(k+1-n)]^T \quad (7.4)$$

where $u(k)$ is the input signal assumed to be bounded and $K-$periodic, $u(k+K) = u(k)$.

For analysis of this implementable algorithm it is convenient to consider its nonimplementable form

$$\hat{\theta}(k+1) = \hat{\theta}(k)-\epsilon\phi(k)(\bar{\phi}^T(k)\hat{\theta}(k+1)-y(k+1)) \quad (7.5)$$

Along with $\bar{\phi}(k)$, we define the "plant regressor"

$$\phi(k) = [y(k) \ y(k-1) \ \cdots \ y(k+1-n)u(k) \ \cdots \ u(k+1-n)]^T \quad (7.6)$$

and say that "$y(k)$ is matchable" if $\hat{\theta}(k)$ can be "tuned" to a constant value θ_* such that

$$y(k+1) = \phi^T(k)\theta_* \ , \quad \forall k \quad (7.7)$$

If $u(k)$ is a persistently exciting (PE) signal for the plant, then the output matchability (7.7) implies the plant-identifier transfer function matchability. However, if the identifier is of lower order, then (7.7) can still be achieved, but only for a severely restricted class of inputs $u(k)$. For example, $u(k)$ can be an input containing an exact minimum number of frequencies to be PE for the identifier and, hence, exciting only a matchable part of the plant.

Local stability properties of all such "tuned regimes" $\{\bar{y}(k) = y(k), \hat{\theta}(k) = \theta_*\}$ are of interest, because they show the tendency of the algorithm to converge to, or move away from, an equilibrium point θ_* in the parameter space $\hat{\theta}$. Local properties are analyzed on a linearized model of the algorithm.

3.7.2 Linearization of the Algorithm

To prepare for our local analysis via linearization, we introduce the variables

$$\eta(k) = \bar{y}(k) - y(k), \quad \theta(k) = \hat{\theta}(k) - \theta_*, \quad \psi(k) = \bar{\phi}(k) - \phi(k) \tag{7.8}$$

representing the deviations around a tuned regime. The equation governing the output error is

$$\eta(k+1) = (\phi(k) + \psi(k))^T \theta(k+1) + \psi^T(k)\theta_* \tag{7.9}$$

where, denoting by a_i^* the tuned value of $\hat{a}_i(k)$,

$$\psi^T(k)\theta_* = -\sum_{i=1}^{n} a_i^* \eta(k+1-i) \tag{7.10}$$

In terms of (7.8) the update law (7.5) is rewritten as

$$\theta(k+1) = \theta(k) - \epsilon(\phi(k) + \psi(k))\eta(k+1) \tag{7.11}$$

It is clear from (7.9)-(7.10) that $\eta(k)$ is the output of the transfer function

$$H(z) = \frac{z^n}{z^n + a_1^* z^{n-1} + \cdots + a_{n-1}^* z + a_n^*} \tag{7.12}$$

when its input is $(\phi(k-1) + \psi(k-1))^T \theta(k)$. The linearization of (7.9)-(7.11) in the neighborhood of a tuned regime is simply performed by considering $\eta(k)$, $\theta(k)$ and $\psi(k)$ as small deviations and neglecting the second order terms $\psi^T(k)\theta(k+1)$ and $\psi(k)\eta(k+1)$. The result is the linear time-varying system

$$\theta(k+1) = \theta(k) - \epsilon\phi(k)\eta(k+1) \tag{7.13}$$

$$\eta(k+1) = -\sum_{i=1}^{n} a_i^* \eta(k+1-i) + \phi^T(k)\theta(k+1) \tag{7.14}$$

Using $\eta(k)$ and its delayed values as the state of $H(z)$,

$$x(k) = [\eta(k) \cdots \eta(k-n+1)]^T \tag{7.15}$$

we rewrite (7.13)-(7.14) as

$$\theta(k+1) = \theta(k) - \epsilon\phi(k)[\phi^T(k)\theta(k+1) + c^T x(k)] \tag{7.16}$$

$$x(k+1) = Ax(k) + b\phi^T(k)\theta(k+1) \tag{7.17}$$

$$x \in \mathbb{R}^n, \quad \theta \in \mathbb{R}^m, \quad m = n+r$$

where

$$A = \begin{bmatrix} -a_1^* & -a_2^* & \cdots & -a_{n_a}^* \\ 1 & 0 & \cdots & 0 \\ \vdots & \ddots & \ddots & \vdots \\ 0 & & \ddots\, 1 & \ddots\, 0 \end{bmatrix}, \quad b = \begin{bmatrix} 1 \\ 0 \\ \vdots \\ 0 \end{bmatrix}, \quad c = \begin{bmatrix} -a_1^* \\ -a_2^* \\ \vdots \\ -a_n^* \end{bmatrix} \tag{7.18}$$

We now apply the state-space decomposition and the averaging analysis.

3.7.3 Decomposition and Averaging

The $L-$transformation of Section 3.3 yields

$$\theta(k+1) = [I - \epsilon g(k,\epsilon)\phi(k)v^T(k,\epsilon)]\theta(k) - \epsilon g(k,\epsilon)\phi(k)c^T\xi(k) \tag{7.19}$$

$$\xi(k+1) = [A + \epsilon g(k,\epsilon)(L(k+1,\epsilon) - b\phi^T(k))\phi(k)c^T]\xi(k) \tag{7.20}$$

where the notation ξ is used instead of z to avoid confusion with the complex variable of the $z-$transform (7.12). Although the application of the stability criteria of Section 5 is straightforward and will be illustrated by an example, several facts specific to this algorithm deserve to be mentioned.

When $H(z)$ is SPR, Goodwin and Sin (1984) prove the convergence of (7.9)-(7.11) using $W(k) = \theta^T(k)\theta(k)$ and showing that its increment $W(k+K-1)-W(k)$ is negative for all k because

$$\frac{1}{K}\sum_{i=k}^{k+K-1} \theta^T(i)\phi(i)\eta(i+1) > 0 \ , \quad \forall k \tag{7.21}$$

To show that our stability criterion in Theorem 3.8 can be interpreted via an approximation of (7.21) we note from (7.13)-(7.17) that

$$\eta(k+1) = c^T x(k) + \phi^T(k)\theta(k+1) = g(k,\epsilon)(\phi^T(k)\theta(k) + c^T x(k)) \tag{7.22}$$

where $g(k,\epsilon) = [1+\epsilon\phi^T(k)\phi(k)]^{-1} = 1-O(\epsilon) < 1$. Substituting

$$c^T x(k) = c^T\xi(k) + c^T L_0(k) + \epsilon c^T L_1(k,\epsilon) \tag{7.23}$$

into (7.22) and, disregarding the exponentially decaying term $\xi(k)$, we obtain

$$\eta(k+1) = [\phi^T(k) + c^T L_0(k)]\theta(k) + O(\epsilon) = v_0^T(k)\theta(k) + O(\epsilon) \tag{7.24}$$

Approximating $\bar{\phi}(k)\eta(k+1)$ by $\phi(k)v_0^T(k)\theta(k)$ reduces (7.21) to

$$\frac{1}{K}\sum_{i=k}^{k+K-1} \theta^T(i)\phi(i)v_0^T(i)\theta(i) > 0 \ , \quad \forall k \tag{7.25}$$

In Theorem 3.8 we have guaranteed the positivity of the sum in (7.25) by requiring that $\bar{R}+\bar{R}^T > 0$, where

$$\bar{R} = \frac{1}{K}\sum_{i=k}^{k+K-1} \phi(i)v_0^T(i) \tag{7.26}$$

(see (5.17)). This is equivalent to taking $W = \theta^T\theta$ as a Lyapunov function for the $\theta-$system (7.19) which, after the $L-$transformation, can be treated separately from the $\xi-$system (7.20). The use of $W = \theta^T\theta$ resulted in our "average SPR" condition of Theorem 3.8, which, in general, is more conservative than the eigenvalue condition of Theorem 3.7. We illustrate this in *Case c* of the example below, where the average SPR condition failed to show the stabilizing effect of adding a d.c. component to the unstable *Case b*. Of course, this effect was shown by the necessary and sufficient condition of Theorem 3.7.

3.7.4 Example

As an illustration we present results of simulation experiments in which the output error algorithm (7.1)-(7.4) was used to estimate the parameters $a_1^* = -1.6$ and $a_2^* = .8$ in the transfer function

$$H(z) = \frac{z}{z^2 - 1.6z + .8} \tag{7.27}$$

Note that $H(z)$ is stable, but *not* SPR. For the input signal

$$u(k) = \sum_{l=-K_1}^{K_2} r_l e^{j\omega_l k} \tag{7.28}$$

where $K_1 \leq 1/2K$, $K_2 \leq 1/2(K-1)$, $K_1+K_2 = K$, the matrix $\bar{R}$ is found to be

$$\bar{R} = \sum_{l=-K_1}^{K_2} |H(e^{j\omega_l})|^2 r_l^2 H(e^{j\omega_l}) \begin{bmatrix} 1 & e^{-j\omega_l} \\ e^{j\omega_l} & 1 \end{bmatrix} \tag{7.29}$$

Case a: Let the input $u(k)$ and the corresponding $\bar{R}$ be

$$u(k) = 0.064\cos\left(\frac{2\pi}{13}k\right), \quad \bar{R} = \begin{bmatrix} 1.02 & 1.98 \\ 1.98 & 1.02 \end{bmatrix} \tag{7.30}$$

where $u(k)$ has period $K = 13$. In this case $\lambda(\bar{R}) = (1.02 \pm j0.58)$ and both Theorems 3.7 and 3.8 predict exponential stability. This is confirmed by simulated trajectories in the $(\hat{a}_1, \hat{a}_2)$ plane shown in Figure 3.8a.

Case b: If instead of (7.30) we use

$$u(k) = 1.2\cos\left(\frac{4\pi}{13}k\right), \quad \bar{R} = \begin{bmatrix} -1.07 & -3.19 \\ 1.98 & -1.07 \end{bmatrix} \tag{7.31}$$

then $\lambda(\bar{R}) = (-1.07 \pm j2.51)$ and Theorem 3.11 predicts instability.

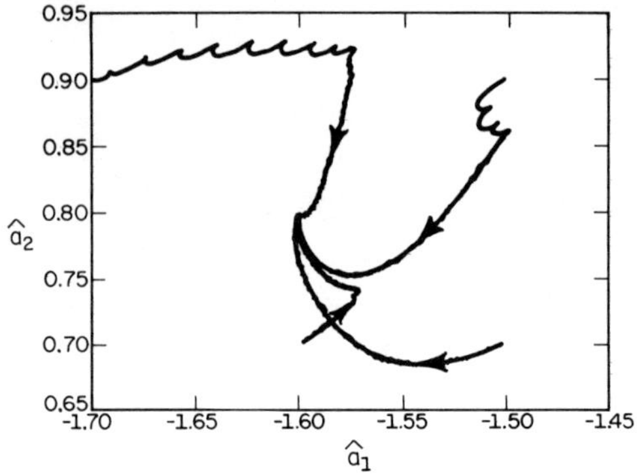

Fig. 3.8a Stable trajectories in Case a satisfying both stability conditions of Theorems 3.7 and 3.8.

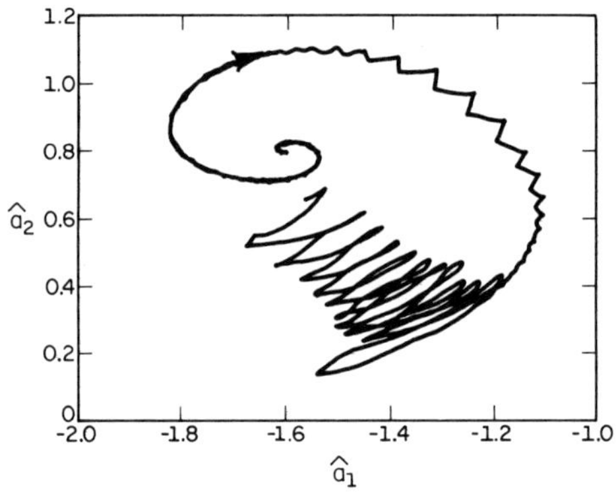

Fig. 3.8b Unstable trajectories in Case b satisfying the instability condition of Theorem 3.6.

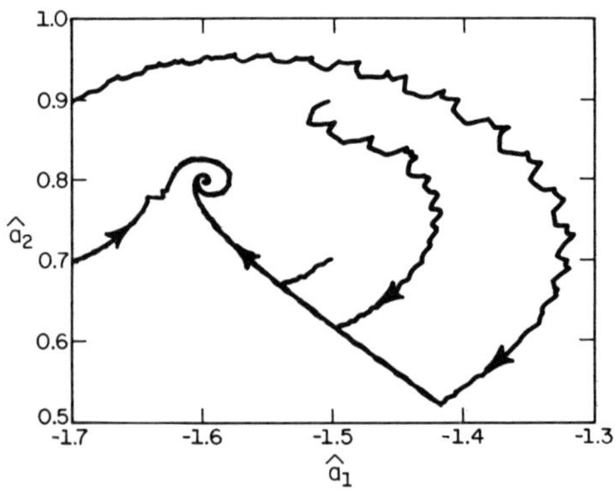

Fig. 3.8c Stable trajectories in Case c satisfying the stability condition of Theorem 3.7, but not the average SPR condition of Theorem 3.8.

A simulated trajectory shown in Figure 3.8b "unwinds," as predicted. However, this local instability does not necessarily imply unboundedness and it appears that the trajectory in Figure 3.8b has a tendency to remain bounded.

Case c: To the input $u(k)$ in (7.31), which led to instability, we now add a d.c. component, namely

$$u(k) = 0.11+1.2\cos\left(\frac{4\pi}{13}k\right), \quad \bar{R} = \begin{bmatrix} .45 & -1.68 \\ 3.49 & .45 \end{bmatrix} \quad (7.32)$$

This final experiment is of interest because the average SPR sufficient condition of Theorem 3.8 is not satisfied, while $\lambda(\bar{R}) = (0.45 \pm jR2.42)$ *satisfies the exponential stability condition of Theorem 3.7. The prediction of exponential stability is confirmed by trajectories in Figure 3.8c. Near the equilibrium they are similar to the trajectories of a linear system with oscillatory but stable eigenvalues.*

Most of the stable simulated trajectories show the characteristic pattern of a fast transient followed by a slow "averaged" behavior. The same pattern, but in the reverse direction of time, is present in unstable trajectories. The local theory predicts the qualitative behavior of the averaged parts of trajectories. When the parameter error is large, the adaptation is fast even for small values of ϵ, and thc assumptions of local theory are not met.

3.8 NOTES AND REFERENCES

Most of the themes and theorems of this chapter evolved from an initial result of Riedle and Kokotovic (1984, 1985b) which delimited the stability-instability boundary for slow adaptation by using an averaging theorem due to Bogoliubov, e.g., Hale (1980), or Miller and Michel (1982). In the periodic case, self-contained derivations of this result were given in Kokotovic, Riedle, and Praly (1985) for the continuous-time case, and in Riedle, Praly, and

Kokotovic (1986) and in Praly and Rhode (1985) for the discrete-time case. These references made use of the L-transformation which is one of the standard singular perturbation tools, e.g., Kokotovic, Khalil, and O'Reilly (1986). The generalization of these periodic results to a wide class of signals, represented in this chapter by sample average Theorems 3.1 and 3.5, is due to Kosut, Anderson, and Mareels (1985). A further result is given by Fu, Bodson, and Sastry (1985) and an application to decentralized adaptive controllers by Ortega and Kelly (1985).

The operator decompositions of Section 3.2 are due to Mareels and Bitmead (1985) and can be used to extend the results of the subsequent sections to the infinite dimensional problems. Examples 3.6.1 and 3.7.4 are discussed in more detail in Riedle and Kokotovic (1985a) and Riedle, Praly, and Kokotovic (1986), respectively.

The averaging approach to adaptive control problems is, of course, much broader than presented in this chapter. Astrom (1983, 1984) has used it for a study of deterministic continuous-time adaptation, while Meerkov (1980) has applied it to develop the so-called vibrational control method. A basic reference on deterministic averaging is Volosov (1962). Averaging methods are more common in stochastic adaptive algorithms, e.g., Ljung (1977), Ljung and Soderstrom (1983), Benveniste (1984). Also related to this chapter are the stochastic averaging results of Bitmead and Boel (1985) and Arnold, Crauerl and Wihstutz (1983).

Chapter 4

STABILITY OF CONTINUOUS-TIME ADAPTIVE SYSTEMS

4.1 INTRODUCTION

In this chapter we study continuous-time adaptive identification and control systems and establish conditions for *global* and *local* stability. As discussed in Chapter 1 the local stability results are obtained by linearization in the neighborhood of a tuned solution. Thus, the local stability properties of the nonlinear (adaptive) system are inherited from the linearized system. In the previous two chapters we presented a detailed study of the stability properties of the linearized system. Here, we draw on these results to provide a qualitative analysis of continuous-time adaptive systems. In Chapter 5 we continue the analysis for the discrete-time case.

Before presenting the local analysis we will first establish the stability properties under ideal conditions of the following well known adaptive identification and control algorithms: equation error parameter estimation, output error parameter estimation, and model reference adaptive control. These will all be cast in the form of the error model developed in Chapter 1, i.e.,

$$\begin{aligned} \theta &= \Omega(\theta_0,\phi,e) \\ e &= e_* - H_{ev}v \\ \phi &= \phi_* - H_{\phi v}v \\ v &= \phi^T\theta \end{aligned} \tag{1.1}$$

where $e(t)$ is the adaptation error, $\phi(t)$ is the regressor, $v(t)$ is the adaptive control error, $\Omega : (\theta_0, \phi, e) \to \theta$ is the parameter adaptive algorithm, and $\theta(t)$ is the adaptive parameter error, given by

$$\theta(t) = \hat{\theta}(t) - \theta_*$$

where $\hat{\theta}(t)$ is the current parameter estimate and θ_* is a constant tuned setting which presumably makes $e_*(t)$ small. Also, H_{ev} and $H_{\phi v}$, as well as $\phi_*(t)$, depend on the choice of θ_*.

Whenever we need to emphasize the θ_*−dependence implicit in (1.1) we will use the explicit notation

$$\begin{aligned} e_*(t) &\equiv e(t, \theta_*) \\ \phi_*(t) &\equiv \phi(t, \theta_*) \\ H_{ev} &\equiv H_{ev}(\theta_*) \\ H_{\phi v} &\equiv H_{\phi v}(\theta_*) \end{aligned} \tag{1.2}$$

In the ideal cases which we will discuss in the next few sections, it is possible to select θ_* such that $e_*(t) = 0$. Once the ideal assumptions are violated, this choice is no longer possible.

Choice of Parameter Adaptive Algorithm: As discussed in Chapter 1, the choice of the parameter adaptive algorithm, i.e., the map $\hat{\Omega} : (\hat{\theta}_0, \phi, e) \to \hat{\theta}$, is usually based on an off-line gradient or Gauss-Newton parameter optimization. Typical algorithms, in continuous-time form, include gradient, least-squares and other variants as described, for example, in Section 1.4.2 of Chapter 1. The analysis of the adaptive error system (1.1) with any of these parameter update algorithms is quite similar. Consequently, we will concentrate on the simple gradient algorithm

$$\dot{\hat{\theta}} = \epsilon \phi e \tag{1.3}$$

although in a few places in the text we will analyze the effect of using the normalized gradient algorithm

$$\dot{\theta} = \frac{\epsilon \phi e}{1+c|\phi|^2} \tag{1.4}$$

Even though the analysis is similar, these algorithms do exhibit differences in performance. The reader is referred to Chapter 5 (Section 1.2) and to Chapter 6 (Section 6.2) for further discussion on these differences.

For easy reference, we write the adaptive error system with the gradient algorithm (1.3) as

$$\begin{aligned} \dot{\theta} &= \epsilon \phi e \\ e &= e_* - H_{ev}(\phi^T \theta) \\ \phi &= \phi_* - H_{\phi v}(\phi^T \theta) \end{aligned} \tag{1.5}$$

If H_{ev} and $H_{\phi v}$ have transfer functions,

$$\begin{aligned} H_{ev}(s) &= d + c^T(sI-A)^{-1}b \\ H_{\phi v}(s) &= D(sI-A)^{-1}b \end{aligned} \tag{1.6}$$

with $A \in \mathbb{R}^{n \times n}$, $D \in \mathbb{R}^{p \times n}$, $d \in \mathbb{R}$, and $b,c \in \mathbb{R}^n$, then we have the state space representation,

$$\begin{aligned} \dot{\theta} &= \epsilon q(t,\theta,z) \\ \dot{z} &= F(\theta)z + G(\theta)\phi_*(t) \end{aligned} \tag{1.7a}$$

where

$$\begin{aligned} q(t,\theta,z) &= (\phi_*(t)-Dz)[e_*(t)-d(\phi_*(t)-Dz)^T\theta-c^Tz] \\ F(\theta) &= A-b\theta^TD \equiv A(\theta_*+\theta) \\ G(\theta) &= b\theta^T \end{aligned} \tag{1.7b}$$

Note also that $\mathrm{Re}\lambda(A(\theta_*)) < 0$ because both $H_{ev}(s,\theta_*)$ and $H_{\phi v}(s,\theta_*)$ are stable.

As we proceed in the next few sections, the reader may find it convenient to refer to Table 4.1. The Table displays the tuned signals (e_*, ϕ_*) and interconnection operators $(H_{ev}, H_{\phi v})$ corresponding to the three adaptive systems we will examine, namely equation error and output error parameter estimation, and model reference adaptive control. The Table shows the case when exact matching is impossible, and uses a shorthand notation which hopefully is clear from the text.

Table 4.1 Error Systems

Equation Error	**Output Error**	**Model Reference Adaptive Control**
	System	
$y = d+Pu$ $e = y-\hat{y},\ \hat{y} = \phi^T\hat{\theta}$ $\phi^T = (Fu, -Fy)$ $\hat{\theta}^T = (\hat{b}, \hat{a}-g)$	$y = d+Pu$ $e = y-\hat{y},\ \hat{y} = \phi^T\hat{\theta}$ $\phi^T = (Fu, -F\hat{y})$ $\hat{\theta}^T = (\hat{b}, \hat{a}-g)$	$y = d+Pu,\ u = -\phi^T\hat{\theta}$ $e = y-y_r,\ y_r = H_r r$ $\phi^T = (Fu, Fy, -r)$ $\hat{\theta}^T = (\hat{a}-g, \hat{b}, \hat{\gamma})$
	Tuned Signals	
$\theta_*^T = (b_*, a_*-g)$ $\phi_*^T = (Fu_*, -Fy_*)$ $u_* = u$ $y_* = y$ $P_* = B_*/A_*$ $e_* = (A_*/G)$ $\cdot[d+(P-P_*)u]$	$\theta_*^T = (b_*, a_*-g)$ $\phi_*^T = (Fu_*, -Fy_*)$ $u_* = u$ $y_* = d+P_*u$ $P_* = B_*/A_*$ $e_* = C[d+(P-P_*)u]$	$\theta_*^T = (a_*-g, b_*, \gamma_*)$ $\phi_*^T = (Fu_*, Fy_*, -r)$ $u_* = -C_*y_*+K_*r$ $y_* = e_*+H_r r$ $C_* = B_*/A_*,$ $K_* = \gamma_*G/A_*$ $e_* = [(1+PC_*)^{-1}PK_*-H_r]r$ $-(1+PC_*)^{-1}d$
	Interconnections	
$H_{ev} = 1$ $H_{\phi v} = 0$	$H_{ev} = CG_*/A_*$ $H_{\phi v}^T = (G/A_*)[0, F]$	$H_{ev} = (1+PC_*)^{-1}PK_*/\gamma_*$ $H_{\phi v}^T = [(H_{ev}/P)F, H_{ev}F, 0]$

4.2 ERROR MODELS: EXACT MATCHING

4.2.1 Equation Error Parameter Estimation

We first consider one of the simplest of adaptive algorithms, namely, equation error parameter estimation. Suppose we have a plant described by

$$y = P_0 u \tag{2.1}$$

where $u(t)$ and $y(t)$ are the input and output measurements, and P_0 represents the plant dynamics. We will assume that P_0 has a rational, strictly proper, transfer function $P_0(s)$, given by

$$P_0(s) = \frac{B_0(s)}{A_0(s)} = \frac{b_o s^m + b_1 s^{m-1} \cdots + b_m}{s^n + a_1 s^{n-1} + \cdots + a_n} \tag{2.2}$$

with $n > m$. As is usual in parameter estimation, we assume that either the plant is stable ($A_0(s)$ is Hurwitz) or else a stabilizing feedback is applied. Thus, the input and output measurements $u(t)$ and $y(t)$ are bounded. Observe that we may also express (2.1) as a *linear regression*,

$$y = \phi^T \theta_0 \tag{2.3}$$

where $\theta_0, \phi(t) \in \mathbb{R}^p, p = n+m+1$, are given by

$$\phi^T = [F_{n-m}u \,, \ldots, F_n u \,, -F_1 y \,, \ldots, -F_n y] \tag{2.4}$$

$$\theta_0^T = [b_o \,, \ldots, b_m \,, a_1 - g_1 \,, \ldots, a_n - g_n] \tag{2.5}$$

and $F_i, i = 1,\ldots,n,$ is a family of stable filters whose transfer functions are

$$F_i(s) = \frac{s^{n-i}}{G(s)} = \frac{s^{n-i}}{s^n + g_1 s^{n-1} + \cdots + g_n} \quad , \quad i = 1,\ldots,n \tag{2.6}$$

Hence, $G(s)$ is Hurwitz. Since the parameters $\theta_0 \in \mathbb{R}^p$ are unknown we consider parameter updates in accordance with the *equation error*,

$$e = y - \phi^T\hat{\theta} \tag{2.7}$$

where $\hat{\theta}(t)$ is an estimate of θ_0 at time t. In this case an *exact matching condition* is satisfied, i.e., $e(t) = 0$ whenever $\hat{\theta}(t) = \theta_0$. Hence, we can choose θ_0 as the tuned parameter setting θ_*.

To analyze the system define the parameter error

$$\theta(t) = \hat{\theta}(t) - \theta_0 \tag{2.8}$$

Thus, combining (2.3)-(2.8) we have

$$e = -\phi^T\theta \tag{2.9}$$

In terms of the adaptive error model (2.1) the tuned signals are

$$e_* = 0 \tag{2.10a}$$

$$\phi_* = \phi \tag{2.10b}$$

and the interconnections are

$$H_{ev} = 1 \tag{2.11a}$$

$$H_{\phi v} = 0 \tag{2.11b}$$

With the gradient algorithm

$$\dot{\hat{\theta}} = \epsilon\phi e$$

the error system becomes

$$\dot{\theta} = -\epsilon\phi\phi^T\theta \tag{2.12}$$

which is a *linear* differential equation, because $\phi_*(t) = \phi(t)$ is dependent only on the input-output data and *not* on the estimate $\hat{\theta}(t)$. Hence, the linear analysis of Chapter 2 (Section 2.3) is applicable and we obtain the following properties:

(P1) $\phi \in L_\infty \Rightarrow \theta \in L_\infty, \dot{\theta} \in L_\infty \cap L_2, e \in L_\infty \cap L_2$

(P2) ϕ uniformly continuous $\Rightarrow \dot{\theta}(t) \to 0, \quad e(t) \to 0$

$$\text{(P3)} \quad \phi \in PE \;\Rightarrow\; \theta(t)\,,\, \dot{\theta}(t)\,,\, e(t) \to 0 \;\text{ exponentially}$$

Property (P1) shows that whenever $\phi(t)$ is bounded we can immediately conclude that $\theta(t)$, $\dot{\theta}(t)$, and $e(t)$ are also bounded. In addition $e(t)$ and $\dot{\theta}(t)$ are in L_2, and, hence, uniform continuity of $\phi(t)$ yields property (P2). Note that the continuity condition is really a minor technical matter, because it will always be satisfied since $u, y \in L_\infty$ and $F_i(s)$, $i = 1,\ldots,n$, are strictly proper and stable.

Property (P3) states that parameter convergence occurs whenever ϕ is persistently exciting, denoted by $\phi \in PE$, i.e., $\exists\ T > 0$ and $\beta \geq \alpha > 0$ such that

$$\beta I_p \;\geq\; \frac{1}{T} \int_s^{s+T} \phi(t)\phi(t)^T dt \;\geq\; \alpha I_p \;\;,\;\; \forall\ s \in \mathbb{R}_+ \qquad (2.13)$$

As shown in Chapter 2 (Section 2.5), $\phi \in PE$ is established if $B_0(s)$, $A_0(s)$ are coprime and if $u(t)$ is *sufficiently rich*, e.g., if $u(t)$ consists of a sum of sinusoids with at least $p/2$ *distinct* frequencies or p spectral lines.

As a matter of interest it also follows that

$$\text{(P4)} \quad \phi \in L_\infty \;\Rightarrow\; \theta \in L_\infty \,,\, e \in L_2$$

Thus, even if $\phi(t)$ is ultimately unbounded, $\theta(t)$ and $e(t)$ can still be well behaved. This property can be regarded as a property of the parameter estimator which motivates its use in adaptive control. By using the normalized gradient algorithm (P4) can be strengthened as follows. We write (1.3) as

$$\dot{\hat{\theta}} \;=\; \epsilon m^2 \phi e \qquad (2.14a)$$

$$m \;=\; (1 + c|\phi|^2)^{-1/2} \;\;,\quad c > 0 \qquad (2.14b)$$

The error system is then

$$\dot{\theta} \;=\; -\, \epsilon m^2 \phi\phi^T \theta \qquad (2.15)$$

Differentiating $|\theta(t)|^2$ along (2.15) and using $e = -\,\phi^T\theta$ gives

$$|\theta(t)| \le |\theta(0)| \tag{2.16a}$$

$$\|(me)_t\|_2 \le |\theta(0)|/(2\epsilon)^{1/2} \tag{2.16b}$$

Now, using the fact that

$$\|(m\phi)_t\|_\infty \le (1/c)^{1/2} \tag{2.17}$$

we also have

$$\begin{aligned} |\dot{\theta}(t)| &\le \epsilon\|(m\phi)_t\|_\infty^2 |\theta(t)| \le (\epsilon/c)|\theta(0)| \\ \|\dot{\theta}\|_{t2} &\le \epsilon\|(m\phi)_t\|_\infty \, \|(me)_t\|_2 \le (\epsilon/2c)^{1/2}|\theta(0)| \\ |m(t)e(t)| &\le \|(m\phi)_t\|_\infty \, |\theta(t)| \le (1/c)^{1/2}|\theta(0)| \end{aligned} \tag{2.18}$$

Hence, property (P4) now reads

$$\text{(P4)}' \quad \phi \in L_\infty \Rightarrow \theta \in L_\infty \,, \quad \dot{\theta} \in L_2 \cap L_\infty$$
$$me \in L_2 \cap L_\infty$$

Comparing (P4)′ to (P4) shows that the principal advantage of normalized gradient to ordinary gradient lies in the fact that $\dot{\theta}$ is in $L_2 \cap L_\infty$ irrespective of any boundedness condition of the regressor $\phi(t)$. When $\phi(t)$ is bounded, say $\|\phi\|_\infty \le a$, then

$$|m(t)e(t)| \ge (1 + ca^2)^{-1/2} \, |e(t)|$$

and it follows from (2.16) that $e \in L_2 \cap L_\infty$. Thus, (P1) becomes

$$\text{(P1)}' \quad \phi \in L_\infty \Rightarrow e \in L_2 \cap L_\infty$$

The remaining properties (P3), (P4) are the same. The fact that normalized gradient (as well as least squares) leads to $\dot{\theta} \in L_2$ has greater significance in proving stability in the case of exact matching for indirect adaptive algorithms, i.e., where control design follows an explicit parameter estimation.

4.2.2 Output Error Parameter Estimation

In this approach the linear regression model is a function of the parameter estimator $\hat{\theta}(t)$. The adaptive algorithm is driven by the *output error* which is defined as

$$e = C(y - \hat{y}) \tag{2.19a}$$

$$\hat{y} = \phi^T\hat{\theta} \tag{2.19b}$$

where C is a stable filter with transfer function $C(s)$. The regressor is now

$$\phi^T = [F_{n-m}u, \ldots, F_n u, -F_1\hat{y}, \ldots, -F_n\hat{y}] \tag{2.20}$$

where again $P_0, F_1, \ldots, F_n$ have transfer functions $P_0(s)$, $F_1(s), \ldots, F_n(s)$ as given by (2.2)-(2.6). Although (2.19) has a similar outward form as (2.7) it is not a linear regression because $\phi(t)$ is directly dependent on $\hat{\theta}(t)$. Thus, there is a parameter feedback loop in the output error, and to distinguish it from (2.7), the term *pseudo-linear regression* is used, (see, e.g., Ljung and Soderstrom (1983)).

The development of an error model is facilitated by defining the *plant regressor*,

$$\phi_0^T = [F_{n-m}u, \ldots, F_n u, -F_1 y, \ldots, -F_n y] \tag{2.21}$$

Hence, the plant is governed by

$$y = \phi_0^T\theta_0 \tag{2.22}$$

with θ_0 given by (2.5). Again using $\theta = \hat{\theta} - \theta_0$, and defining the unfiltered output error

$$\eta = y - \hat{y} \tag{2.23}$$

we have

$$\begin{aligned} \eta &= (\phi_0 - \phi)^T\theta_0 - \phi^T\theta \\ &= -[0,\ldots,0,F_1\eta,\ldots,F_n\eta]\theta_0 - \phi^T\theta \\ &= -[(a_1-g_1)F_1 + \cdots + (a_n-g_n)F_n]\eta - \phi^T\theta \\ &= \frac{G-A_0}{G}\eta - \phi^T\theta \end{aligned}$$

Hence,

$$\eta = -\frac{G}{A_0}\phi^T\theta \tag{2.24}$$

As a result, the tuned signals in the error model are

$$e_* = 0 \tag{2.25a}$$

$$\phi_* = \phi_0 \tag{2.25b}$$

with tuned interconnections

$$H_{ev} = CG/A_0 \tag{2.26a}$$

$$H_{\phi v} = (G/A_0)[0,\ldots,0, F_1,\ldots, F_n]^T \tag{2.26b}$$

As in the case of equation error parameter estimation, the tuned parameter θ_* is θ_0, which means that an exact matching condition holds, i.e., $e_*(t) = 0$.

The major differences between output error and equation error can be seen by comparing the interconnection operators. In the equation error case we have $H_{ev} = 1$, $H_{\phi v} = 0$, whereas here $H_{ev} \neq 1$ and $H_{\phi v} \neq 0$. The latter means that $\phi(t)$ is dependent on $\hat{\theta}(t)$, i.e., there is a feedback mechanism present and so the error system is not linear. Consequently, it is not immediately apparent that the linear analysis of Chapter 2 is applicable. However, as will be seen subsequently, $H_{ev}(s) \in \mathrm{SPR}$ renders harmless the nonlinearity, and thus yields the same results as in the linear system.

In the output error case, if, for example, $C(s)$ is strictly proper, then $H_{ev}(s)$ as given by (2.26) is SPR, provided that

(1) $C(s)G(s)/A_0(s)$ is stable.

(2) $\exists$ constant $\rho > 0$ such that

$$\mathrm{Re}[C(j\omega)G(j\omega)/A_0(j\omega)] \geq \rho|C(j\omega)G(j\omega)/A_0(j\omega)|^2, \ \forall\ \omega \in \mathbb{R}_+ \tag{2.27}$$

Notice also that the elements of $H_{\phi v}(s)$ are all stable and strictly proper. This property as well as the SPR property of $H_{ev}(s)$ will appear again in the case of model reference adaptive control, where the form of the error system is the same. Therefore, we next develop the error model for adaptive control, and then present the stability analysis for the exact matching case with $H_{ev}(s) \in$ SPR and $e_*(t) = 0$.

4.2.3 Model Reference Adaptive Control

Known Plant: We first consider non-adaptively controlling the plant,

$$y = P_0 u \tag{2.28}$$

where as before (2.2) P_0 has transfer function

$$P_0(s) = \frac{B_0(s)}{A_0(s)} = \frac{b_o s^m + b_1 s^{m-1} + \cdots + b_m}{s^n + a_1 s^{n-1} + \cdots + a_n} \tag{2.29}$$

with $m < n$. We assume that the plant is known in the sense that the parameters $\{b_o, \ldots, b_m, a_1, \ldots, a_n\}$ are known. The plant is to be controlled so that its response matches closely that of a known stable *reference model,*

$$y_{\text{ref}} = H_{\text{ref}} r \tag{2.30}$$

where H_{ref} has the transfer function,

$$H_{\text{ref}}(s) = \frac{B_r(s)}{A_r(s)} = \frac{b_0^r s^m + b_1^r s^{m-1} + \cdots + b_m^r}{s^n + a_1^r s^{n-1} + \cdots + a_n^r} \tag{2.31}$$

and $r(t)$ is the reference model input. We will consider the linear control form

$$u = C_{ur} r - C_{uy} y \tag{2.32}$$

where C_{ur}, C_{uy} have proper transfer functions $C_{ur}(s)$ and $C_{uy}(s)$, respectively. One approach to obtain exact matching of the transfer function $r \to y$ to $H_{\text{ref}}(s)$ is as follows: consider the polynomials

$$T(s) = s^{n_T} + t_1 s^{n_T - 1} + \cdots + t_{n_T}$$

$$S(s) = s^{n_T} + s_1 s^{n_T-1} + \cdots + s_{n_T} \tag{2.33}$$

$$R(s) = r_1 s^{n-1} + \cdots + r_n$$

which satisfy

$$T(s)A_r(s) = R(s) + A_0(s)S(s) \tag{2.34}$$

If $T(s), A_r(s)$, and $A_0(s)$ are known, then there are unique polynomials $R(s)$ and $S(s)$ which solve (2.34) provided that

$$n_T \geq n - m - 1 \tag{2.35}$$

If the controller is selected as

$$C_{uy}(s) = \frac{R(s)}{B_0(s)S(s)} \tag{2.36a}$$

$$C_{ur}(s) = \frac{B_r(s)T(s)}{B_0(s)S(s)} \tag{2.36b}$$

then exact matching of the reference model is obtained and closed loop internal stability is secured provided that $B_0(s)$ and $T(s)$ are Hurwitz. This latter condition restricts the plant to be minimum phase, but it is necessary to obtain exact matching of an *arbitrary* stable reference model, and thus avoid unstable pole-zero cancellations. Putting all the assumptions together we have

(A1) $P_0(s)$ is known and given by (2.29)

(A2) $H_r(s)$ is known and given by (2.31)

(A3) $T(s)$ is Hurwitz

(A4) $B_0(s)$ is Hurwitz

Note that the polynomial $T(s)$ can be thought of as representing observer dynamics. Further details on the observer representation are provided in Chapter 5, Section 2.1.

Now, define the polynomial

$$G(s) = s^{n_G} + g_1 s^{n_G-1} + \cdots + g_{n_G} \tag{2.37}$$

with

$$n_G = n_T + m \geq n - 1 \tag{2.38}$$

because $n_T \geq n-m-1$. The control $u(t)$ given by (2.32), (2.36) can be expressed equivalently as

$$u = \frac{G-\gamma_0 B_0 S}{G} u - \frac{\gamma_0 R}{G} y + \frac{\gamma_0 B_r T}{G} r \tag{2.39}$$

where

$$\gamma_0 = 1/b_o \tag{2.40}$$

Hence, u has the *linear control form*

$$u = - \phi^T \theta_0 \tag{2.41}$$

where θ_0, $\phi(t) \in \mathbb{R}^p$, $p = n_G+n+1$, are given by

$$\phi^T = \left[(F_1,\ldots,F_{n_G})u,\ (F_{n_G-n+1},\ldots,F_{n_G})y,\ -\frac{B_r T}{G} r \right] \tag{2.42}$$

$$\theta_0^T = [\alpha_1,\ldots,\alpha_{n_G},\beta_1,\ldots,\beta_n,\gamma_0] \tag{2.43}$$

with α the coefficients of $\gamma_0 B_0(s)S(s) - G(s)$, and β the coefficients of $\gamma_0 R(s)$ the filters are now defined by

$$F_i(s) = s^{n_G-i}/G(s)\ , \quad i = 1,\ldots,n_G \tag{2.44}$$

It follows from (2.38) that the number of control parameters p is such that

$$p = n_T + n + m + 1 \geq 2n \tag{2.45}$$

Observe also that the control parameter vector θ_0 is an implicit function of the plant parameters $\{b_o,\ldots,b_m,a_1,\ldots,a_n\}$, the exact relationship defined by (2.4). Thus, the plant is no longer being explicitly identified as in equation error and output error parameter estimation. In this case, we can view (2.41) as the model for identification. With this in mind, we now consider the case of adaptive control.

Adaptive Model Reference Control: We now turn to the case in which the coefficients in $B_0(s)$, $A_0(s)$ are *unknown*, but their order, n and m, is known. The linear control form of (2.41) suggests the *adaptive linear control*

$$u = -\phi^T\hat{\theta} \tag{2.46}$$

with $\phi(t)$ from (2.42) and where $\hat{\theta}(t)$ is an estimate of θ_0. Suppose also that we choose the adaptation error as the tracking error,

$$e = y - y_r \tag{2.47}$$

If the parameter error is defined by

$$\theta(t) = \hat{\theta}(t) - \theta_* , \quad \theta_* = \theta_0$$

with θ_0 from (2.43), then by using (2.39) we obtain the adaptive control signal as

$$u = -\phi^T\theta_0 - \phi^T\theta \tag{2.48a}$$

$$= \frac{G - \gamma_0 B_0 S}{G} u - \frac{\gamma_0 R}{G} y + \frac{\gamma_0 B_r T}{G} r - v \tag{2.48b}$$

where

$$v = \phi^T\theta \tag{2.49}$$

Hence, using (2.36) we have

$$u = -\frac{R}{B_0 S} y + \frac{B_r T}{B_0 S} r - \frac{G}{\gamma_0 B_0 S} v \tag{2.50a}$$

$$= -C_{uy} y + C_{ur} r - \frac{G}{\gamma_0 B_0 S} v \tag{2.50b}$$

Referring to the adaptive error system (1.1), the tuned signals are

$$e_* = 0 \tag{2.51a}$$

$$\phi_*^T = [(H_r/P_0)(F_1,\ldots,F_{n_G}),\ H_r(F_{n_G-n+1},\ldots,F_{n_G}),\ -(B_r T/G)]r \tag{2.51b}$$

and a little calculation gives the tuned interconnections

$$H_{ev} = \frac{b_o G}{TA_r} \tag{2.52a}$$

$$H^T_{\phi v} = [(H_{ev}/P_0)(F_1,\ldots,F_{n_G}),\, H_{ev}(F_{n_G-n+1},\ldots,F_{n_G}),0] \tag{2.52b}$$

Since $n > m$ and $n_G = n_T + m$, it follows that $H_{ev}(s)$ and all the terms in $H_{\phi v}(s)$ are strictly proper. Moreover, with $G(s)$ and $B_0(s)$ Hurwitz, they are also stable as required by the definition of a tuned setting.

If the gradient algorithm is applied, then the adaptive error system is

$$\dot{\theta} = -\epsilon\phi H_{ev}(\phi^T\theta) \tag{2.53a}$$

$$\phi = \phi_* - H_{\phi v}(\phi^T\theta) \tag{2.53b}$$

which has the same form as the error system for output error parameter estimation. Hence, to obtain global stability results, we again require that $H_{ev}(s) \in$ SPR. For example, suppose we choose

$$G(s) = T(s)B_r(s) \tag{2.54}$$

Then,

$$H_{ev}(s) = b_o H_r(s) \tag{2.55}$$

(This choice is not unique but is convenient to illustrate the analysis.) Since $H_r(s)$ is strictly proper ($n > m$), $H_{ev}(s) \in$ SPR if and only if there is a constant $\rho > 0$ such that

$$\mathrm{Re}[b_o H_r(j\omega)] \geq \rho b_0^2 |H_r(j\omega)|^2 \quad , \quad \forall\ \omega \geq 0 \tag{2.56}$$

Thus, it follows that a necessary condition for $b_o H_r(s) \in$ SPR is that $n-m = 1$, i.e., the plant $P_0(s)$ *must have relative degree of one.* This is obviously a serious restriction and we will return to the implications of loss of SPR later on. Note that our algorithm is designed for a plant with relative degree one, and using a different algorithm would be a reasonable consideration. However, loss of SPR is more fundamental, arising from the impossibility of achieving an exact matching condition due to unstructured model errors.

A second necessary condition for (2.56) is that the sign of the high frequency gain of $b_o H_r(s)$ be positive. Since $n-m = 1$, this essentially requires a priori knowledge about the sign of b_o, which is the high frequency gain of $P_0(s)$. The relative degree condition, however, is much more restrictive and, as will be shown in the next section, *cannot be eliminated under a passivity analysis.*

Summarizing the assumptions so far, we have:

(A1) Plant order n is known

(A2) Sign of b_o is known

(A3) $P_0(s)$ has relative degree one

(A4) $P_0(s)$ is minimum phase

(A5) $b_o H_r(s)$ is SPR and strictly proper

Assumption (A5) can be modified so that $b_o H_r(s)$ is SPR, proper, but not strictly proper. The modification is accomplished by inserting filters in the regressor signals and removing them in the parameter update algorithm; see, e.g., Narendra, Lin, and Valavani (1980) or Egardt (1979). The stability analysis, however, remains essentially unaltered.

Remarks on Estimating the High Frequency Gain: The need to know the sign of b_o is a result of using the simple adaptive linear control form (2.41). This can be overcome by using methods which provide an estimate of b_o as part of the update algorithm. In Chapter 5, Section 2, we utilize an algorithm which does provide this estimate. As noted there, except for minor modifications, the stability analysis is similar. The major difference, however, is that when estimating b_o directly, the regressor consists of plant inputs and outputs, whereas in the linear control form (2.41), the regressor also contains reference signal. The implications of this distinction will be considered in Chapter 6.

4.3 GLOBAL STABILITY: PASSIVITY ANALYSIS

4.3.1 The Main Theorem

The global stability assumptions and properties are gathered together and stated in the following theorem.

Theorem 4.1: Global Stability

Consider the system

$$\begin{aligned} \dot{\theta} &= \epsilon \phi e \ , \qquad \theta(0) \in \mathbb{R}^p \\ e &= e_* - H_{ev}(\phi^T \theta) \\ \phi &= \phi_* - H_{\phi v}(\phi^T \theta) \end{aligned} \tag{3.1}$$

where

(A1) $H_{ev}(s)$ is SPR

(A2) $H_{\phi v}(s)$ is stable and strictly proper

Under these conditions:

(i) $\phi_* \in L_\infty$ $\Rightarrow$ $\theta \in L_\infty$, $\dot{\theta} \in L_2 \cap L_\infty$, $\dot{\theta}(t) \to 0$ (3.2)
$e_*(t) \to 0$ exp

$$e, \dot{e} \in L_2 \cap L_\infty \ , \ e(t) \to 0 \tag{3.3}$$

$$\phi - \phi_* \in L_2 \cap L_\infty \ , \ \phi(t) - \phi_*(t) \to 0 \tag{3.4}$$

(ii) $$\phi_* \in PE \;\Rightarrow\; \theta(t)\,,\ \dot{\theta}(t)\,,\ e(t),\ \text{and}\ \phi(t)-\phi_*(t) \quad \text{all} \to 0 \ \text{exponentially fast} \tag{3.5}$$

Remark: Recall from Chapter 2 that when $H_{ev}(s)$ is SPR and strictly proper, $\phi_* \in PE$ is defined by (2.3.18), and when $H_{ev}(s)$ is SPR, proper, but not strictly proper, $\phi_* \in PE$ is given by (2.3.17). These PE conditions are equivalent whenever $\phi_*(t)$ has bounded derivative except for a countable number of points of fixed minimum separation.

Proof

We use the state-space representation of (3.1), i.e.,

$$\begin{aligned} \dot{\theta} &= \epsilon\phi e \quad , \quad e = -c^T z - d\phi^T\theta \\ \dot{z} &= Az + b\phi^T\theta \quad , \quad \phi = \phi_* - Dz \end{aligned} \tag{3.6}$$

where $H_{ev}(s) = d + c^T(sI-A)^{-1}b$ and $H_{\phi v}(s) = D(sI-A)^{-1}b$. Since $H_{ev}(s) \in SPR$, $\exists\ \mu > 0,\ d \geq 0,\ A \in \mathbb{R}^{n\times n}$, and $b, c, q \in \mathbb{R}^n$ such that (see Chapter 2, Lemma 2.6)

$$A + A^T = -\mu I - qq^T \tag{3.7a}$$

$$b = c - (2d)^{1/2} q \tag{3.7b}$$

Differentiating the function

$$V = \frac{1}{\epsilon}\,|\theta|^2 + |z|^2 \tag{3.8}$$

along (3.6), and using (3.7) results in

$$\dot{V} = -\mu|z|^2 - ((2d)^{1/2}\phi^T\theta + q^T z)^2 \tag{3.9}$$

Integrating from $[0,t]$ we obtain $\forall\ t \geq 0$,

$$\max\left\{ \frac{1}{\epsilon}\,|\theta(t)|^2\,,\ |z(t)|^2\,,\ \mu\|z_t\|_2^2 \right\} \leq V(0) \tag{3.10}$$

In our operator formulation of the error model, the expression $H_{ev}(\phi^T\theta)$ is defined on $t \in \mathbb{R}_+$. Hence, the system states $z(t)$ start at $z(0) = 0$. Thus, $V(0) = \frac{1}{\epsilon}|\theta(0)|^2$, and we have $\forall\ t \geq 0$,

$$\begin{aligned} |\theta(t)| &\leq |\theta(0)| \\ |z(t)| &\leq |\theta(0)|/(\epsilon)^{1/2} \\ \|z_t\|_2 &\leq |\theta(0)|/(\mu\epsilon)^{1/2} \end{aligned} \tag{3.11}$$

Allowing $z(0) \neq 0$ is equivalent to $e_*(t) \neq 0$, but where $e_*(t) \to 0$ exponentially. Thus, in either case, we establish immediately that $\theta \in L_\infty$ and $z \in L_\infty \cap L_2$. Note in particular that when $e_*(t) = 0$ (equivalently $z(0) = 0$),

$$\|(\phi-\phi_*)_t\|_2 = \|(Dz)_t\|_2 \tag{3.12a}$$

$$\leq |D|\cdot|\theta(0)|/(\mu\epsilon)^{1/2} \tag{3.12b}$$

and that

$$|\phi(t)-\phi_*(t)| \leq |D|\cdot|\theta(0)|/(\epsilon)^{1/2} \tag{3.13}$$

Hence, upper bounds on $\phi(t)-\phi_*(t)$ are dependent only on the initial parameter estimate.

Since $H_{ev}(s) \in \text{SPR}$, either $d > 0$ or $d = 0$. When $d = 0$, by chasing the signals around the loop we obtain the following series of implications: $e \in L_2 \cap L_\infty$, $\phi-\phi_* \in L_2 \cap L_\infty \Rightarrow \phi \in L_\infty \Rightarrow \dot{z} \in L_\infty \Rightarrow z(t) \to 0 \Rightarrow e(t) \to 0$, $\phi(t)-\phi_*(t) \to 0$, $\dot{\theta}(t) \to 0$. This establishes (3.2)-(3.4), and (3.5) follows from the results of Chapter 2 on PE systems. When $d > 0$, a similar series of implications yield the same results.

□

4.3.2 Discussion of Limitations

The conditions of the Global Stability Theorem impose a PE condition on the tuned regressor $\phi_*(t)$. In the case of equation error parameter estimation, $\phi_*(t)$ consists of filtered plant input/output measurements (2.4). Thus, provided $\{B_0(s), A_0(s)\}$ are coprime, PE follows from a richness condition on the plant input $u(t)$, e.g., $u(t)$

contains a sufficient number of spectral lines. In output error parameter estimation and model reference adaptive control, the regressor $\phi(t)$ depends on the parameter estimate $\hat{\theta}(t)$; thus, $\phi(t) \neq \phi_*(t)$. However, in the course of the proof we were able to show that $\phi(t)$ inherits the PE condition of $\phi_*(t)$ *because of exact matching*, i.e., whenever $H_{ev}(s) \in \text{SPR}$ and either $e_*(t) = 0$ or $e_*(t) \to 0$ exponentially, we have $\phi(t) \to \phi_*(t)$ as $t \to \infty$. Consequently, in output error parameter estimation, the tuned regressor is given by (2.21), and hence, provided $\{B_0(s), A_0(s)\}$ is coprime, a sufficient richness condition on the plant input $u(t)$ provides for $\phi_*(t) \in PE$. In model reference adaptive control, with the filter choice (2.56), the tuned regressor is

$$\phi_* = [(H_r/P_0)(F_1,\ldots,F_{n_G}), H_r(F_1,\ldots,F_n), -1]r \qquad (3.14)$$

Appealing to Theorem 2.7, it follows that $\phi_* \in PE$ if both pairs $\{B_r(s), A_r(s)\}$ and $\{B_0(s), A_0(s)\}$ are coprime, and if $r(t)$ is sufficiently rich, e.g., if $r(t)$ is a sum of at least $p/2$ distinct sinusoids or else contains at least p spectral lines.

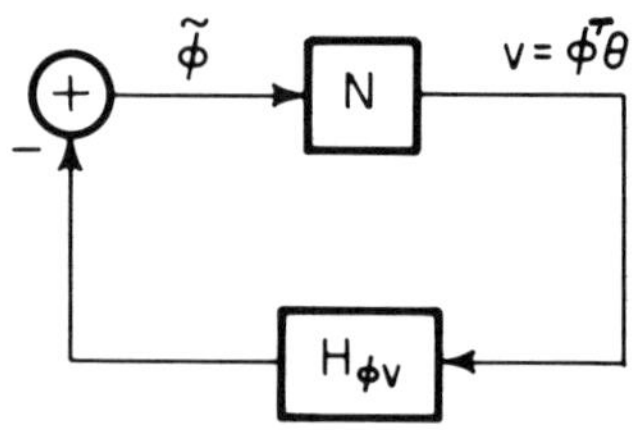

Fig. 4.1 Adaptive system: regressor loop.

Another interesting interpretation of the global theory can be obtained by viewing the adaptive error system (1.1) as the feedback interconnection of two operators as depicted in Figure 4.1. The forward-path operator $N: \tilde{\phi} \to v$ is described by

$$\phi = \phi_* + \tilde{\phi} \qquad (3.15a)$$

$$\dot{\theta} = \epsilon[\phi e_* - \phi H_{ev}(\phi^T\theta)] \quad , \quad \theta(0) = \theta_0 \tag{3.15b}$$

$$v = \phi^T\theta \tag{3.15c}$$

The return signal $\tilde{\phi}$ is obtained from the feedback path

$$\tilde{\phi} = -H_{\phi v}(\phi^T\theta) \tag{3.16}$$

Observe that (3.15b) is in effect the *linear adaptive system* that we have studied in previous chapters where we consider $\phi(t)$ as a given signal.

Clearly $\tilde{\phi}$ is the amount by which the regressor $\phi(t)$ differs from the tuned regressor $\phi_*(t)$. We can also view the feedback loop as the incremental mapping $T: \tilde{\phi} \to \tilde{\psi}$ given by

$$\begin{aligned} \tilde{\psi} &= -H_{\phi v}(\phi^T\theta) \\ \phi &= \phi_* + \tilde{\phi} \\ \dot{\theta} &= \epsilon[\phi e_* - H_{ev}(\phi^T\theta)] \quad , \quad \theta(0) = \theta_0 \end{aligned} \tag{3.17}$$

Observe that the error system is then completely equivalent to the single operator expression

$$\tilde{\phi} = T\tilde{\phi} \tag{3.18}$$

We will refer to T as the *incremental regressor loop-gain* operator. Now, we apply small gain theory: the theory asserts (see Chapter 2, Section 2.2) that if, for some $p \in [1,\infty]$, the L_p-gain of T is less than one, then the system is L_p-stable. This is equivalent to showing that

$$\gamma_p(T) = \gamma_p(H_{\phi v}N) < 1 \tag{3.19}$$

with N defined by (3.15). From our previous analysis of the ideal case, when $H_{ev}(s) \in$ SPR and $e_*(t) = 0$ we have (see (3.12))

$$\|\tilde{\psi}\|_2 \leq |D|\cdot|\theta(0)|/(\mu\epsilon)^{1/2} \tag{3.20}$$

Since $\|\tilde{\psi}\|_2$ does *not* depend on $\tilde{\phi}$, it follows that $\gamma_2(T) = 0$. Thus, under ideal conditions the *incremental regressor loop-gain is zero!*

This result is really at the core of the global theory. Essentially, the nonlinear structure of the adaptive system is transparent under the zero loop-gain phenomena. Unfortunately, the zero loop-gain property is lost under non-ideal conditions, e.g., when exact matching is impossible, and, thus, $H_{ev}(s)$ is not SPR and $e_*(t)$ is not zero.

These disclaimers lead us away from the global approach of Theorem 4.1 to the establishment of "local" stability results as discussed in Chapter 1. The problem then is to determine pragmatic conditions that maintain ϕ within a satisfactory neighborhood of some persistently exciting function, and hence preserve exponential stability.

Before embarking in this direction, let us look a little more deeply into the effect of unmodelled dynamics, i.e., loss of exact matching. We will provide some examples in Section 4.4.4 to illustrate that various types of instabilities may arise. First, however, we develop the error models for our representative adaptive systems when exact matching is impossible.

4.4 ERROR MODELS: EXACT MATCHING IMPOSSIBLE

4.4.1 Parameter Estimation

Suppose that the plant is *actually* described by

$$y = Pu + d \tag{4.1}$$

where P is linear with transfer function $P(s)$ and $d(t)$ is a disturbance in L_∞. We will assume that $P(s)$ is more complicated than $P_0(s)$ in (2.2), i.e., there is no set of parameters $\{a_1, \ldots, a_n, b_o, \ldots, b_m\}$ as defined there such that $P(s) = P_0(s)$. This is the case, for example, whenever $P(s)$ is of higher order than $P_0(s)$. Moreover, we also assume that $d(t)$ is an arbitrary, but bounded, L_∞-function. Hence, there is no $\theta \in \mathbb{R}^p$ which can zero either the equation error

$$e = y - \phi^T\theta \tag{4.2}$$

with $\phi(t)$ from (2.4), or the output error

$$e = C(y - \phi^T\theta) \tag{4.3}$$

with $\phi(t)$ from (2.20). If we use the same algorithms as before but with the plant given by (4.1), then the tuned signals and systems in the adaptive error model undergo some modifications. It is convenient to introduce a *tuned parameter* setting $\theta_* \in \mathbb{R}^p$, $p = n+m+1$ defined by

$$\theta_* = [b_0^*, \ldots, b_m^*, a_1^* - g_1, \ldots, a_n^* - g_n]^T \tag{4.4}$$

where the constants $\{b_0^*, \ldots, b_m^*, a_1^*, \ldots, a_n^*\}$ appear as coefficients in an associated *tuned plant* P_*, whose transfer function is

$$P_*(s) = \frac{b_0^* s^m + \cdots + b_m^*}{s^n + a_1^* s^{n-1} + \cdots + a_n^*} = \frac{B_*(s)}{A_*(s)} \tag{4.5}$$

The tuned plant can be interpreted as a best fit choice given a particular data set $\{u(\cdot), y(\cdot)\}$. For example, θ_* could arise from minimization of a mean-square error criterion.

For equation error we then have the tuned signals

$$\begin{aligned} e_* &= (A_*/G)[d + (P-P_*)u] \\ \phi_*^T &= [F_{n-m}u, \ldots, F_n u, -F_1 y, \ldots, F_n y] \end{aligned} \tag{4.6}$$

and tuned interconnections

$$H_{ev} = 1 \tag{4.7a}$$

$$H_{\phi v} = 0 \tag{4.7b}$$

Similarly, for output error we have the tuned signals

$$\begin{aligned} e_* &= C[d + (P - P_*)u] \\ \phi_*^T &= [F_{n-m}u, \dots, F_n u, -F_1 y_*, \dots -F_n y_*] \\ y_* &= d + P_* u \end{aligned} \tag{4.8}$$

and tuned interconnections

$$H_{ev} = CG/A_* \tag{4.9a}$$

$$H_{\phi v} = (G/A_*)[0, \dots, 0, F_1, \dots, F_n]^T \tag{4.9b}$$

The pattern of tuned signals and systems when exact matching is not possible reveals some interesting differences between equation error and output error parameter estimations.

For equation error we retain the SPR property in $H_{ev}(s)$; in fact, $H_{ev}(s) = 1$. Thus, the error system is

$$\dot{\theta} = \epsilon[\phi_* e_* - \phi_* \phi_*^T \theta] \tag{4.10}$$

which is a linear *inhomogeneous* system. Consequently, the linear analysis from Chapter 2 is directly applicable, and for equation error we obtain the following *global stability* result directly from Theorem 2.2.

Lemma 4.1: Equation Error Stability

For an equation error parameter estimation with the error model given by (4.10), if $\phi_* \in PE$ and $e_* \in L_\infty$ then $\theta \in L_\infty^p$.

□

Since $d(t)$ is an arbitrary L_∞-function, there are implicit spectral and/or magnitude restrictions on $d(t)$ to maintain $\phi_* \in PE$. For example, we can write

$$\phi_*(t) = \bar{\phi}_*(t) + \tilde{\phi}_*(t) \tag{4.11}$$

where $\tilde{\phi}_*(t) = 0$ whenever $d(t) = 0$. Suppose that $\bar{\phi}_* \in PE$ as defined by (2.13) with constants $T > 0$ and $\beta \geq \alpha > 0$. Since

$\phi_* \in PE$ requires that

$$\lambda_{\min}\{ \frac{1}{T} \int_s^{s+T} \phi_*(t)\phi_*(t)^T dt \} > 0 \quad , \forall\, s \in \mathbb{R}_+$$

it follows by substituting (4.11) into (2.13) that

$$\alpha > \|\tilde{\phi}_*\|_\infty (2\|\bar{\phi}_*\|_\infty + \|\tilde{\phi}_*\|_\infty) \qquad (4.12)$$

is a sufficient condition. Further algebraic manipulations on (4.12) would translate into an upper bound on $\|d\|_\infty$ to insure that $\phi_* \in PE$.

If $e_*(t)$ is small enough -- which is also affected by $d(t)$ -- then $\theta(t)$ will also ultimately be small *from any starting guess* $\theta(0)$. Thus, as $t \to \infty$, $\hat{\theta}(t)$ approaches an $O(\|e_*\|_\infty)$ neighborhood of the tuned setting θ_*. So, in equation error, under a PE condition on the regressor, we always get a bounded estimate.

One of the problems with equation error can be seen from the equation for e_*, viz. (4.6). The "filter" A_*/G is dependent on θ_*, thus, direct modification of its bandwidth is only indirectly controllable through $G(s)$. Compare this to the expression for $e_*(t)$ from output error in (4.8), where the filter $C(s)$ is completely controlled by the designer. Hence, it is possible in output error to be less sensitive to noise and model error than in equation error. However, in output error the regressor ϕ is partly determined by a feedback mechanism; therefore, the interconnection $H_{ev}(s)$ given by (4.9); cannot be guaranteed to be SPR, i.e., it is dependent on the choice of θ_*.

Any loss of SPR in output error precludes the use of the Global Stability Theorem. As loss of SPR can also occur in adaptive control, we postpone introducing our local stability results. These will be developed for the generic error model with $e_*(t) \neq 0$ and $H_{ev}(s)$ not SPR. We next develop the error model for adaptive control when exact matching is impossible.

4.4.2 Adaptive Control

As in Section 4.4.1, suppose that the actual plant to be controlled is

$$y = Pu + d \tag{4.13}$$

where the transfer function $P(s)$ is more complicated than $P_0(s)$, i.e., $P(s)$ is higher order than $P_0(s)$, and $d(t)$ is an arbitrary disturbance in L_∞. We will express the error system for this case in two forms. The first involves the tuned plant as defined in (4.5), and the second involves a tuned controller.

Error Model with Tuned Plant: As in (4.5) let P_* denote the *tuned plant,*

$$P_*(s) = \frac{b_0^* s^{n-1} + \cdots + b_n^*}{s^n + a_1^* s^{n-1} + \cdots + a_n^*} = \frac{B_*(s)}{A_*(s)} \tag{4.14}$$

The tuned plant is presumed to satisfy the following ideal assumptions:

(A1) Minimum phase ($B_*(s)$ is Hurwitz)

(A2) Relative degree one ($b_0^* \neq 0$)

(A3) Known sign of high frequency gain

Since P_* is *not* the actual plant, these assumptions do not apply to the actual plant P. Hence, the actual plant P can be non-minimum phase, have any relative degree greater than one, and its high frequency gain can be unknown.

Associated with $A_*(s)$ we also have the tuned polynomials $R_*(s)$, $S_*(s)$ which uniquely solve (2.34), i.e.,

$$T(s)A_r(s) = R_*(s) + A_*(s)S_*(s) \tag{4.15}$$

Suppose we use the same algorithm as in Section 4.2.3, that is,

$$\dot{\hat{\theta}} = \epsilon\phi e \ , \quad e = y - H_{\text{ref}} r \tag{4.16}$$

with

$$u = -\phi^T\hat{\theta} \tag{4.17a}$$

$$\phi^T = [F_1 u, \ldots, F_{n_G} u, F_{n_G-n+1} y, \ldots, F_{n_G} y, -r] \tag{4.17b}$$

where

$$\begin{aligned} F_i(s) &= s^{n_G - i}/G(s) \\ G(s) &= B_r(s)T(s) \\ n_G &= m + n_T \end{aligned} \tag{4.18}$$

The tuned signals and systems are expressed more succinctly by introducing

$$M_* = \frac{R_*}{TA_r} \tag{4.19a}$$

$$\tilde{P} = (P - P_*)/P_* \tag{4.19b}$$

The tuned signals are then

$$\begin{aligned} e_* &= (1+\tilde{P}M_*)^{-1}\left[\tilde{P}(1-M_*)H_r r + \frac{A_* S_*}{A_r T}\, d\right] \\ u_* &= (1+\tilde{P}M_*)^{-1}[(b_0^* H_r/P_*)r - (M_*/P_*)d] \\ \phi_* &= [F_1 u_*, \ldots, F_{n_G} u_*, F_{n_G-n+1}(y_r + e_*), \ldots, F_{n_G}(y_r + e_*), -r] \end{aligned} \tag{4.20}$$

The interconnections are

$$\begin{aligned} H_{ev} &= (1+\tilde{P}M_*)^{-1}(1+\tilde{P})b_0^* H_r \\ H_{\phi v}^T &= [(1+\tilde{P}M_*)^{-1}(b_0^* H_r/P_*)(F_1, \ldots, F_{n_G}), \\ &\quad (1+\tilde{P}M_*)^{-1}(1+\tilde{P})b_0^* H_r(F_1, \ldots, F_n), 0] \end{aligned} \tag{4.21}$$

When $\tilde{P} = 0, P = P_* = P_0,$ and we return to the case of exact matching. Since $B_*(s)$ is Hurwitz, the interconnections are stable and the tuned signals are bounded if and only if $\tilde{P}$ and $(1+\tilde{P}M_*)^{-1}$ are stable. For example, if $\tilde{P}$ is stable, $(1+\tilde{P}M_*)^{-1}$ is stable if

$$|\tilde{P}(j\omega)| < 1/|M_*(j\omega)| \ , \ \forall \ \omega \in \mathbb{R}_+ \tag{4.22}$$

since M_* is stable. From the definition (4.19), $\tilde{P}(s)$ is also a function of θ_*. Thus, (4.22) is a constraint on choices of θ_* which will stabilize the tuned system. One may ask the question as to the possibility of preserving SPR despite model error. Inequality (4.22) implies the possibility, but unfortunately, it does not take very much model error to upset the SPR condition. One way to see this is to express H_{ev} in terms of the plant P directly, as we now show.

Error Model with Tuned Controller: Let $u_*(t)$ denote the tuned control signal given by (4.17) with $\hat{\theta}(t) = \theta_*$, i.e.,

$$u_* = \phi_*^T\theta_* \tag{4.23}$$

where

$$\phi_*^T = [F_1u_*,\ldots,F_{n_G}u_*,F_{n_G-n+1}y_*,\ldots,F_{n_G}y_*,-r] \tag{4.24a}$$

$$\theta_*^T = [\alpha_1-g_1,\ldots,\alpha_{n_G}-g_{n_G},\beta_1,\ldots,\beta_n,\gamma_0] \tag{4.24b}$$

Thus,

$$u_* = C_{ur}^*r - C_{uy}^*y_* \tag{4.25}$$

where

$$C_{uy}^*(s) = \frac{N_*(s)}{D_*(s)} = \frac{\beta_1^*s^{n-1} + \cdots + \beta_n^*}{s^{n_G}+\alpha_1^*s^{n_G-1} + \cdots + \alpha_{n_G}^*} \tag{4.26}$$

$$C_{ur}^*(s) = \frac{\gamma_0^*G(s)}{D_*(s)} = \frac{\gamma_0^*B_r(s)T(s)}{D_*(s)} \tag{4.27}$$

Suppose the actual plant has the strictly proper transfer function,

$$P(s) = \frac{B(s)}{A(s)} \tag{4.28}$$

where $B(s)$ and $A(s)$ are polynomials such that

$$\deg[A(s)] > n$$

Then in terms of the tuned controller parameters, the tuned signals are

$$
\begin{aligned}
e_* &= \left(\frac{\gamma_0^* BG}{AD_* + BN_*} - \frac{B_r}{A_r} \right) r - \left(\frac{AD_*}{AD_* + BN_*} \right) d \\
y_* &= e_* + \frac{B_r}{A_r} r \\
u_* &= \left(\frac{\gamma_0^* AG}{AD_* + BN_*} \right) r - \left(\frac{AN_*}{AD_* + BN_*} \right) d
\end{aligned}
\tag{4.29}
$$

Also, the interconnections are

$$
H_{ev} = \frac{BG}{AD_* + BN_*} \tag{4.30}
$$

$$
H_{\phi v} = \begin{bmatrix} \dfrac{AG}{AD_* + BN_*} (F_1, \ldots, F_{n_G})^T \\[2ex] \dfrac{BG}{AD_* + BN_*} (F_{n_G+n-1}, \ldots, F_{n_G})^T \\[2ex] 0 \end{bmatrix} \tag{4.31}
$$

Clearly, by definition of the tuned setting, we require that

$$
A(s)D_*(s) + B(s)N_*(s) \quad \text{is Hurwitz} \tag{4.32}
$$

4.4.3 Loss of SPR Condition

Since $H_{ev}(s) \in$ SPR requires that rel deg$[H_{ev}(s)] \leq 1$, it follows from (4.30) that rel deg$[P(s)] \leq 1$. In other words, if rel deg$[P(s)] \geq 2$, then $H_{ev}(s)$ cannot be SPR for any choice of θ_*. As a result, neglecting to model even the most benign of high frequency phenomena will violate the SPR condition. For example, suppose

$$
P(s) = \frac{ab}{(s+a)(s+b)} P_*(s) \tag{4.33}
$$

with $a > b$ and b much greater than the bandwidth of $P_*(s)$. Then rel deg$[P(s)]$ = rel deg$[P_*(s)]+2 > 1$ and the SPR condition on $H_{ev}(s)$ cannot hold. Another informative interpretation of loss of SPR can be obtained by writing

$$H_{ev}(s) = \bar{H}_{ev}(s) + \tilde{H}_{ev}(s) \tag{4.34}$$

where $\bar{H}_{ev}(s)$ is the (nominal) transfer function when there is no uncertainty, i.e., when exact matching is possible. One can also take the view that $\tilde{H}_{ev}(s)$ represents the *unmatchable* part of the closed-loop dynamics. For example, from (4.29) and (4.30) we can write (explicitly as a function of θ_*)

$$H_{ev}(s,\theta_*) = \frac{1}{\gamma_0^*} H_{yr}(s,\theta_*) \tag{4.35}$$

where $H_{yr}(s,\theta_*)$ is the tuned closed-loop system transfer function from $r \to y$, which should be close to the reference model $H_{\text{ref}}(s)$. In terms of the notation in (4.34) we have

$$\bar{H}_{ev}(s,\theta_*) = \frac{1}{\gamma_0^*} H_{\text{ref}}(s) \tag{4.36}$$

$$\tilde{H}_{ev}(s,\theta_*) = \frac{1}{\gamma_0^*} [H_{yr}(s,\theta_*) - H_{\text{ref}}(s)] \tag{4.37}$$

We can also write

$$H_{ev}(s,\theta_*) = [H_{yr}(s,\theta_*)/H_{\text{ref}}(s)]\bar{H}_{ev}(s,\theta_*) \tag{4.38}$$

The ratio H_{yr}/H_{ref}, which ideally is one, also provides a useful interpretation of the impossibility of exactly matching $H_{yr}(s,\theta_*)$ with $H_{\text{ref}}(s)$ for any $\theta_* \in \mathbb{R}^p$.

The loss of SPR as well as the presence of bounded disturbances not only preclude exact matching, and hence, preclude the global results of Theorem 4.1, they also open the door to various types of instabilities. These are described next.

4.4.4 Instability Mechanisms

In this section we demonstrate instabilities which arise when there is a loss of SPR and/or a loss of PE. In the cases shown, the onset of the instability is triggered by *parameter drift*. For example, if the parameters drift out of the constant parameter stability set, it is likely that the then exponentially growing transients will significantly disrupt performance -- even if saturation is not reached.

Consider the plant to be controlled described by

$$y(t) = d(t) + P(s,\mu)u(t) \tag{4.39a}$$

where

$$d(t) = \frac{1}{s-a}\, w(t) \tag{4.39b}$$

$$P(s,\mu) = \frac{1-\mu s}{(s-a)(1+\mu s)} \tag{4.39c}$$

The constant a is unknown, $\mu > 0$ is a small parasitic time constant, and $w(t)$ is a bounded disturbance. We consider an adaptive regulator based on the plant $P(s,0)$, i.e.,

$$u(t) = -\hat{\theta}(t)y(t) \tag{4.40}$$

$$\dot{\hat{\theta}}(t) = \epsilon y(t)^2 \quad , \ \epsilon > 0 \tag{4.41}$$

Note that $e(t) = \phi(t) = y(t)$ in this simple case. In the disturbance-free case $(w(t) = 0)$ with no parasitics $(\mu = 0)$, Theorem 4.1 asserts that $\hat{\theta}(t)$ is bounded and $y(t) \to 0$ as $t \to \infty$. This can be shown using

$$V(y,\hat{\theta}) = \frac{1}{2}\, y^2 + \frac{1}{2\epsilon}\, (\hat{\theta}-\theta_*)^2 \tag{4.42}$$

as a Lyapunov function for (4.39)-(4.41) when $\mu = 0$ and $w(t) = 0$. The same function shows that for any bounded $w(t) \neq 0$, the output $y(t)$ remains bounded, because

$$\dot{V}(y,\hat{\theta}) \leq -|y|[(\theta_*-a)|y|-|w|] \tag{4.43}$$

Moreover, if $w(t) \to 0$, the regulation of $y(t)$ to zero can still be preserved. For example, if $\epsilon = 1/4$, $\hat{\theta}(0) = 1$, $y(0) = 1$, and

$$w(t) = (1+t)^{-1/8}[1-a(1+t)^{-1/4} - 3/8(1+t)^{-5/4}] \quad (4.44)$$

then

$$y(t) = (1+t)^{-3/8} \quad (4.45a)$$

$$\hat{\theta}(t) = (1+t)^{1/4} \quad (4.45b)$$

Hence, although the output is regulated, i.e.,

$$y(t) \to 0 \quad \text{as} \quad t \to \infty$$

the adaptive parameter $\hat{\theta}(t)$ drifts to infinity, i.e.,

$$\hat{\theta}(t) \to \infty \quad \text{as} \quad t \to \infty \,.$$

In this example we have

$$H_{ev}(s,\theta_*) = [1+\theta_* P(s,0)]^{-1} P(s,0) \quad (4.46a)$$

$$= \frac{1}{s+\theta_*-a} \quad (4.46b)$$

Hence, $H_{ev}(s,\theta_*)$ is SPR for $\theta_* > a$. However, $\phi(t) = y(t)$ is not PE since $y(t) \to 0$ as $t \to \infty$. As a result *parameter drift* can occur with loss of PE whenever disturbances are present, ***even though the SPR condition holds***.

When $\mu > 0$ the SPR condition is lost *no matter how small μ is*, i.e.,

$$H_{ev}(s,\theta_*) = [1+\theta_* P(s,\mu)]^{-1} P(s,\mu)$$

$$= \frac{1-\mu s}{\mu s^2+(1-(a+\theta_*)\mu)s+\theta_*-a} \quad (4.47)$$

Moreover, the closed-loop tuned system is stable if and only if

$$\frac{1}{\mu} - a > \theta_* > a \quad (4.48)$$

Now, suppose the adaptive regulator (4.41) is applied to (4.39) and there are no disturbances present, i.e., $d(t) = 0$. Then the resulting system in state-form is

$$\begin{aligned} \dot{y} &= (a+\hat{\theta})y + z \\ \mu\dot{z} &= -a - 2\hat{\theta}y \\ \dot{\hat{\theta}} &= \epsilon y^2 \end{aligned} \tag{4.49}$$

This system has the equilibrium

$$\begin{aligned} y &= 0 \\ z &= 0 \\ \hat{\theta} &= \theta_* \end{aligned} \tag{4.50}$$

which is stable if and only if (4.48) holds. Since $\dot{\hat{\theta}} \geq 0$, it follows that

$$\hat{\theta}(0) > \frac{1}{\mu} - a \;\Rightarrow\; \hat{\theta}(t) > \frac{1}{\mu} - a \;, \quad \forall\, t \geq 0$$

Hence, from this initial value $\hat{\theta}(0)$ the equilibrium (4.50) cannot be reached. Moreover, with $\hat{\theta}(t) > 1/\mu - a$, the system is unstable even if the adaptation loop is disconnected, i.e., $\epsilon = 0$. This phenomenon, well understood in terms of classical feedback theory, is referred to as a *linear instability*. Observe also that in this case we preserve PE but lose SPR.

The main message of these two examples is this: whenever exact matching is lost, a parameter drifting may occur which signifies the onset of negative effects. Most of our examples of adaptive control systems illustrate this characteristic drifting behavior. The parameters undergo a fast transient followed by a steady slow drifting. To be more precise, we are concerned with a *slow* parameter drifting which occurs in the neighborhood in parameter space corresponding to a small tuned error signal. The problem then is to *stop the drift*. This is accomplished in the linear case (Chapters 2 and 3) by satisfying a signal dependent positivity condition. In the next section we show how the same condition arises in the local analysis of the (nonlinear) adaptive system.

4.5 ROBUSTNESS OF SLOW ADAPTATION: AVERAGING ANALYSIS

In this section we will establish conditions under which the qualitative properties of the drifting phenomena can be predicted under slow adaptation. Our analysis is local and based on the classical methods of linearization and averaging for nonlinear systems. The results to be presented in Section 4.5.1 pertain to systems with almost periodic signals and are somewhat self-contained, although they do connect with the ideas of averaging and linearization presented in Chapters 3 and 1, respectively. In Section 4.5.2 to follow, we will extend these results to account for signals which do not necessarily have a uniform average. These latter results arise directly from the linearization and averaging ideas previously presented.

4.5.1 Nonlinear Averaging: Almost Periodic Signals

We will present the method of averaging for almost periodic systems by considering the gradient algorithm together with the linear adaptive system (1.5). Recall the state-space description of the adaptive error system for this case (1.7), i.e.,

$$\dot{\theta} = \epsilon q(t,z) \tag{5.1a}$$

$$\dot{z} = F(\theta)z + G(\theta)\phi_*(t) \tag{5.1b}$$

with

$$\begin{aligned} q(t,z) &= \phi_*(t)e_*(t) - (\phi_*(t)c^T + e_*(t)D)z + c^T z D z \\ F(\theta) &= A(\theta_*) - b\theta^T D = A(\theta_* + \theta) \qquad (5.2) \\ G(\theta) &= b\theta^T \end{aligned}$$

Also recall that $\theta(t)$ is the parameter error defined by

$$\theta(t) = \hat{\theta}(t) - \theta_*$$

where $\hat{\theta}(t)$ is the current estimate of θ_* which is a tuned setting in

$\mathbb{R}^p$. The objective of adaptation, as espoused by the philosophy in Chapter 1, is to adjust $\hat{\theta}(t)$ so as to arrive in a sufficiently small neighborhood of θ_*, the allowed size of the neighborhood being determined from performance demands. In error space this means that as $t \to \infty$, $\theta(t)$ should be small. Although a small neighborhood of θ_* is certainly the designer's choice when the plant is known, the question remains as to what is the algorithm's choice. That is, does the algorithm converge to a set which includes θ_* and is it small enough to guarantee performance? Clearly, this convergent set is a function of the tuned signals $\phi_*(t)$, $e_*(t)$ and initial conditions $\hat{\theta}(0)$. At a minimum we would like to *sculpture* $\phi_*(t)$, $e_*(t)$ so that if $\hat{\theta}(0)$ starts near θ_*, then it stays near θ_*. This is the question we will examine now using an averaging analysis.

What we first do is define a family of state trajectories corresponding to a *fixed parameter error* $\theta \in \mathbb{R}^p$. A tuned setting, $\theta = 0$, obviously generates one of these trajectories. Let $\bar{z}(t,\theta)$ denote the state $z(t)$ where $\theta(t)$ is *frozen* at some value $\theta \in \mathbb{R}^p$. The system which generates $\bar{z}(t,\theta)$ is referred to as the *frozen parameter error system*. Thus, for each fixed $\theta \in \mathbb{R}^p$, $\bar{z}(t,\theta)$ is the solution of

$$\frac{\partial \bar{z}}{\partial t} = F(\theta)\bar{z} + G(\theta)\phi_*(t) \tag{5.3}$$

Thus, $\bar{z}(t,\theta)$ is given explicitly by

$$\bar{z}(t,\theta) = \int_{-\infty}^{t} \exp\{(t-\tau)F(\theta)\}G(\theta)\phi_*(\tau)d\tau \tag{5.4}$$

Note that initial conditions are accounted for in (5.4) by integrating from the infinite past to the present. Now introduce the error state

$$\eta(t) = z(t) - \bar{z}(t,\theta(t)) \tag{5.5}$$

where $\bar{z}(t,\theta(t))$ is the output of the operator defined implicitly by (5.4) with input θ replaced by $\theta(t)$. The magnitude of $\eta(t)$ reflects the amount by which the system state $z(t)$ is near to a system with parameters frozen at $\theta(t)$. Using (5.3)-(5.5), we can transform (5.1) from (θ,z) coordinates to (θ,η) coordinates. In (θ,η) coordinates the adaptive error system (5.1) becomes

$$\dot{\theta} = \epsilon q(t,\bar{z}(t,\theta) + \eta) \tag{5.6a}$$

$$\dot{\eta} = F(\theta)\eta - \epsilon\bar{\psi}(t,\theta)q(t,\bar{z}(t,\theta)+\eta) \tag{5.6b}$$

where $\bar{\psi}(t,\theta)$ is the matrix function defined by

$$\bar{\psi}(t,\theta) = \partial\bar{z}(t,\theta)/\partial\theta \tag{5.7}$$

Hence, $\bar{\psi}(t,\theta)$ reflects the *frozen state sensitivities* with respect to θ. Observe also that for fixed $\theta \in \mathbb{R}^p$, $\bar{\psi}(t,\theta)$ satisfies

$$\frac{\partial\bar{\psi}}{\partial t} = F(\theta)\,\bar{\psi} + \frac{\partial}{\partial\theta}\,[F(\theta)\bar{z}+G(\theta)\phi_*(t)] \tag{5.8}$$

Other sensitivities can also be computed. For example, let $\bar{e}(t,\theta)$ and $\bar{\phi}(t,\theta)$ denote the frozen parameter error and regressor, respectively. Thus, for fixed $\theta \in \mathbb{R}^p$, we have

$$\bar{e} = e_* - c^T\bar{z} \tag{5.9a}$$

$$\bar{\phi} = \phi_* - D\bar{z} \tag{5.9b}$$

which together with (5.6) and (5.8) gives

$$\frac{\partial\bar{\psi}}{\partial t} = (A(\theta_*)-b\theta^T D)\bar{\psi} + b\bar{\phi}^T \tag{5.10}$$

Recall that $H_{ev}(s,\theta_*) = c^T(sI-A(\theta_*))^{-1}b$ and $H_{\phi v}(s,\theta_*) = D(sI-A(\theta_*))^{-1}b$. Hence, using $\bar{z}(t,0) = 0$ from (5.4), we have the sensitivities,

$$\left(\frac{\partial\bar{e}}{\partial\theta}\right)\Big|_{\theta=0} = -H_{ev}(\theta_*)\phi_*^T \tag{5.11a}$$

$$\left(\frac{\partial\bar{\phi}}{\partial\theta}\right)\Big|_{\theta=0} = -H_{\phi v}(\theta_*)\phi_*^T \tag{5.11b}$$

These relations will be useful later to provide insight into the meaning of the tuned setting θ_*.

It is of primary interest to examine the behavior of (5.6)-(5.8) in the neighborhood of $(\theta,\eta) = (0,0)$. Linearizing (5.6)-(5.8) about $(\theta,\eta) = (0,0)$ gives

$$\dot{\theta} = \epsilon[\phi_* e_* - Q_*(\bar{\psi}_0\theta + \eta)] \tag{5.12a}$$

$$\dot{\eta} = A(\theta_*)\eta - \epsilon\bar{\psi}_0[\phi_* e_* - Q_*(\bar{\psi}_0\theta + \eta)] \tag{5.12b}$$

where

$$Q_* = \phi_* c^T + e_* D \tag{5.12c}$$

and $\bar{\psi}_0(t) = \bar{\psi}(t,0)$ satisfies

$$\dot{\bar{\psi}}_0 = A(\theta_*)\bar{\psi}_0 + b\phi_*^T \tag{5.12d}$$

Observe that

$$\bar{\psi}(t,0) = L_0(t) \tag{5.13}$$

where $L_0(t)$ is the matrix used in the time-scale transformation defined by (3.3.11). Hence, if the adaptation is slow, meaning that $\epsilon > 0$ is small, then the transformation $(\theta,z) \rightarrow (\theta,\eta)$ is also a *time-scale transformation* essentially decoupling the "fast" states $\eta(t)$ from the "slow" states $\theta(t)$. In fact, (5.12) with $e_*(t) = 0$ is precisely the system described by (3.3.56). Hence, the time-scale decomposition of Chapter 3 is obtained as a linearization of (5.6) about $(\theta,\eta) = (0,0)$ with $e_*(t) = 0$. At this point we can analyze the local stability of (5.6) by one of two means: (1) linearization using the averaging results of Chapter 3 and the Total Stability Theorem of Chapter 1, or (2) direct nonlinear averaging. In this subsection we proceed with nonlinear averaging and in the next subsection use linearization and total stability.

We remark that an exact time-scale transformation can be obtained by introducing the *integral manifold* of (5.1). On the manifold the system state $z(t)$ is a function of $\theta(t)$, e.g., $z(t) = h(t,\theta(t))$. For small $\epsilon > 0$, the *manifold function* $h(t,\theta)$ is near the frozen system state $\bar{z}(t,\theta)$. Moreover, it is shown in Riedle and Kokotovic (1985) that the time-scale transformations $(\theta,z) \to (\theta,\eta)$ given here and the $L-$transformation of (3.3.11) are approximating the exact time-scale transformation induced by the manifold function. For our purposes it is sufficient to consider the transformation $(\theta,z) \to (\theta,\eta)$.

Suppose that $\hat{\theta}(t)$ is near θ_* and, under slow adaptation, $\dot{\hat{\theta}}(t)$ is small. In error terms, $\theta(t)$ is near zero and $\dot{\theta}(t)$ is small. Thus, we are motivated to approximate (5.6) locally near $\theta = 0$ by

$$\dot{\theta} = \epsilon q(t,\bar{z}(t,\theta)+\eta)$$

$$\dot{\eta} = A(\theta_*)\eta = F(0)\eta$$

Since $\mathrm{Re}\lambda[A(\theta_*)] < 0$ by definition of the tuned setting, it follows that for small ϵ, $\eta(t) \to 0$ quite rapidly with respect to changes in $\theta(t)$, and hence, $\eta(t)$ has little effect on $\theta(t)$. Moreover, slow adaptation will also average out transients. Thus, if $\hat{\theta}(t)$ is near θ_* and if the system state $z(t)$ is near the frozen state $\bar{z}(t,0)$, then $\theta(t)$ behaves on average like the solution to,

$$\dot{\theta} = \epsilon q_{avg}(\theta) \tag{5.14}$$

where

$$q_{avg}(\theta) = \lim_{T\to\infty} \frac{1}{T} \int_0^T q(t,\bar{z}(t,\theta))dt \tag{5.15}$$

provided the limit exists. Observe that

$$q_{avg}(\theta_e) = 0 \tag{5.16}$$

defines the equilibrium solutions $\theta_e \in \mathbb{R}^p$ of (5.14). Assuming there exists at least one solution, the exponentially stable equilibria for small ϵ, are those such that

$$\mathrm{Re}\,\lambda[\partial q_{avg}(\theta_e)/\partial\theta] < 0 \tag{5.17}$$

If no solution $\theta_e \in \mathbb{R}^p$ of (5.16) exists, then it is possible that the adaptive system will have a finite-escape time, or other erratic behavior. We will assume here that our adaptive system is well enough constructed so that (5.16) can be solved. Notice that we have not required that $\theta_e = 0$ solve (5.16). Hence, our approximate decoupled system is more accurately described by

$$\dot{\theta} = \epsilon q(t,\bar{z}(t,\theta)+\eta) \tag{5.18a}$$

$$\dot{\eta} = F(\theta_e)\eta = A(\theta_* + \theta_e)\eta \tag{5.18b}$$

We can support this approximation by appealing to the following theorem which is a modification of Theorem 3.4 in Hale (1980), which is the classical two-time scale nonlinear averaging result, valid for systems of the form

$$\begin{aligned} \dot{\theta} &= \epsilon f(t,\theta,\eta) \\ \dot{\eta} &= B\eta + \epsilon g(t,\theta,\eta) \end{aligned}$$

where B is a constant matrix. Comparing this system to the adaptive system (5.6), B is replaced by the matrix function $F(\theta)$, and hence, the modification. Although many such variants have appeared in the standard literature, e.g., Volosov (1962), we utilize a theorem statement which follows Hale (1980), but is more like that in Bodson et al. (1985), since the latter is developed specifically for adaptive systems. We state the result here without proof. Following the theorem we will discuss the meaning of θ_e, particularly the condition $\theta_e = 0$, i.e., $q_{avg}(0) = 0$.

Theorem 4.2: Nonlinear Averaging

Consider the adaptive error system (5.6), i.e.,

$$\begin{aligned} \dot{\theta} &= \epsilon q(t,\bar{z}(t,\theta)+\eta) \\ \dot{\eta} &= F(\theta)\eta - \epsilon\bar{\psi}(t,\theta)q(t,\bar{z}(t,\theta)+\eta) \end{aligned}$$

Suppose $q(t,\bar{z}(t,\theta)+\eta)$, $F(\theta)\eta$, and $\bar{\psi}(t,\theta)$ are all continuous in $(t,\theta,\eta) \in \mathbb{R}_+ \times \mathbb{R}^p \times \mathbb{R}^n$, almost periodic in t uniformly with respect to (θ,η) in compact sets, and have continuous first derivatives with respect to (θ,η). Define the functions $q_{avg} : \mathbb{R}^p \to \mathbb{R}^p$ and $R : \mathbb{R}^p \to \mathbb{R}^{p\times p}$ by

$$q_{avg}(\theta) = \lim_{T\to\infty} \frac{1}{T} \int_0^T q(t,\bar{z}(t,\theta))dt \tag{5.19}$$

$$R(\theta) = \partial q_{avg}(\theta)/\partial\theta \tag{5.20}$$

If there is a $\theta_e \in \mathbb{R}^p$ satisfying

$$q_{avg}(\theta_e) = 0 \tag{5.21}$$

for which $\mathrm{Re}\,\lambda[R(\theta_e)] \neq 0$ and $\mathrm{Re}\,\lambda[F(\theta_e)] \neq 0$ then $\exists\ \epsilon_0 > 0$ such that $\forall\ \epsilon \in [0,\epsilon_0]$:

(i) $\exists$ unique almost periodic functions $\bar{\theta}(t,\epsilon)$ and $\bar{\eta}(t,\epsilon)$ satisfying (5.6), and as $\epsilon \to 0$, $\bar{\theta}(t,\epsilon) \to \theta_e$, $\bar{\eta}(t,\epsilon) \to 0$.

(ii) $\max_i\{\mathrm{Re}\,\lambda_i[R(\theta_e)],\ \mathrm{Re}\lambda_i[F(\theta_e)]\} < 0 \Rightarrow \bar{\theta}(t,\epsilon),\ \bar{\eta}(t,\epsilon)$ is u.a.s.

(iii) $\max_i\{\mathrm{Re}\,\lambda_i[R(\theta_e)],\ \mathrm{Re}\lambda_i[F(\theta_e)]\} > 0 \Rightarrow \bar{\theta}(t,\epsilon),\ \bar{\eta}(t,\epsilon)$ is unstable.

□

From the form of (5.1)-(5.2), the conditions in the theorem enumerated prior to (5.19) are all satisfied provided that $e_*(t)$ and $\phi_*(t)$ are almost periodic functions. Since the tuned system is a stable LTI system, it follows that almost periodic solutions $e_*(t)$, $\phi_*(t)$ will exist provided that the reference and disturbance inputs are almost periodic. This, then, is the implied assumption required by the theorem.

Subject to this assumption, let us concentrate our interpretive remarks on the meaning of the equilibrium solution θ_e of (5.21). For convenience, let $\mathrm{avg}\{\cdot\}$ denote averaging with respect to time. Thus, using (5.9) in (5.19) we have,

$$q_{avg}(\theta) = \mathrm{avg}\{\bar{\phi}(\cdot,\theta)\bar{e}(\cdot,\theta)\} \tag{5.22}$$

Recall that in our error notation $\theta(t) = \hat{\theta}(t) - \theta_*$. Hence, in (5.22) we can replace θ with $\hat{\theta} - \theta_*$ where $\hat{\theta}$ now denotes a constant free parameter vector representing the *actual* adjustable parameters. Thus, if θ_* were an equilibrium solution then (5.21) is equivalent to

$$q_{avg}(0) = \text{avg}\{\phi_* \, e_*\} = 0 \tag{5.23}$$

On the other hand, even if θ_* were not an equilibrium, then we can use (5.23) to generate a candidate tuned setting. To be more explicit we can write

$$\phi_*(t) \equiv \phi(t, \theta_*) \tag{5.24a}$$

$$e_*(t) \equiv e(t, \theta_*) \tag{5.24b}$$

Hence, a *candidate tuned setting,* denoted by θ_a, is any solution of

$$q_{avg}(0) = \text{avg}\{\phi(\cdot, \theta_a) e(\cdot, \theta_a)\} = 0 \tag{5.25}$$

where $\phi(t, \hat{\theta})$ and $e(t, \hat{\theta})$ are the regressor and error, respectively, associated with some fixed setting $\hat{\theta}(t) = \hat{\theta}$. (These are to be distinguished from $\overline{\phi}(t, \theta)$ and $\overline{e}(t, \theta)$ which are functions of the fixed parameter error $\theta = \hat{\theta} - \theta_*$.)

Using this definition together with the definition of $F(\theta)$ and $A(\theta_*)$ in (5.2), the stability conditions (ii), (iii) now read:

(ii)′ $\max_i\{\text{Re}\,\lambda_i[R(0)], \text{Re}\lambda_i[A(\theta_a)]\} < 0 \Rightarrow$ u.a.s.

(iii)′ $\max_i\{\text{Re}\,\lambda_i[R(0)], \text{Re}\lambda_i[A(\theta_a)]\} > 0 \Rightarrow$ unstable.

One can view θ_a as the algorithm's choice of a candidate tuned setting, which is not necessarily the user's choice. We may, after computing θ_a, decide that in fact it is a tuned setting. On the other hand, it may not be a tuned setting at all. The latter situation will occur if the signals which are present *during adaptation* are substantially different than those present during the control application. In the extreme, it is even possible that $\text{Re}\lambda[A(\theta_a)] > 0$, which is clearly unstable, and hence, undesirable. We will say more on these issues in Chapter 6, where we also provide further examples and justification for these results which lead ultimately to a design guide.

Let us now compute $R(0) = \partial q_{avg}(0)/\partial\theta$. Using the sensitivity expressions in (5.9)-(5.11), and the notation $(\cdot)_a$ rather than $(\cdot)_*$ to emphasize the origins of the candidate tuned setting θ_a, we have

$$\begin{aligned} R(0) &= \frac{\partial}{\partial\theta}\,\mathrm{avg}\{\bar{e}\bar{\phi}\}|_{\theta=0} \\ &= \mathrm{avg}\left\{e_a\left(\frac{\partial\bar{\phi}}{\partial\theta}\right)|_{\theta=0}+\phi_a\left(\frac{\partial\bar{e}}{\partial\theta}\right)|_{\theta=0}\right\} \\ &= -\,\mathrm{avg}\{e_aH_{\phi v}(\theta_a)\phi_a+\phi_aH_{ev}(\theta_a)\phi_a^T\} \end{aligned} \tag{5.26}$$

Hence, from (ii)$'$, for sufficiently small $\epsilon > 0$, there is a u.a.s. solution of (5.6) if

$$\mathrm{Re}\lambda\{\mathrm{avg}[\phi_aH_{ev}(\theta_a)\phi_a^T + e_aH_{\phi v}(\theta_a)\phi_a^T]\} > 0 \tag{5.27}$$

If $e_a(t)$ is sufficiently small, then it is sufficient to insure that

$$\mathrm{Re}\lambda\{\mathrm{avg}[\phi_aH_{ev}(\theta_a)\phi_a^T]\} > 0 \tag{5.28}$$

This is precisely the signal dependent positivity condition arrived at in Chapter 3 for the linearized adaptive system; see (3.5.15).

Another informative interpretation of $q_{avg}(0) = 0$ is obtained as follows. Consider the $\theta-$dependent average-square-error,

$$J(\theta) = \frac{1}{2}\,\mathrm{avg}\{\bar{e}^2\} \tag{5.29}$$

Let $J_\theta(\theta) = \partial J(\theta)/\partial\theta$. Hence, from (5.11) we have

$$\begin{aligned} J_\theta(0) &= \mathrm{avg}\left\{\bar{e}\left(\frac{\partial\bar{e}}{\partial\theta}\right)\right\}|_{\theta=0} \\ &= -\,\mathrm{avg}\{e_aH_{ev}(\theta_a)\phi_a^T\} \end{aligned} \tag{5.30}$$

Using the fact that $\mathrm{avg}(e_a\phi_a) = 0$ we have

$$J_\theta(0) \;=\; \text{avg}\{e_a[1 - H_{ev}(\theta_a)]\phi_a^T\} \tag{5.31}$$

Similarly, using the sensitivity expressions in (5.11), we obtain the second order partials of (5.29) as given by

$$\begin{aligned} J_{\theta\theta}(0) \;&=\; \text{avg}\{[H_{ev}(\theta_a)\phi_a][H_{ev}(\theta_a)\phi_a]^T\} + O(\|e_a\|_\infty) \\ &=\; \text{avg}\{\phi_a H_{ev}(\theta_a)\phi_a^T + [(H_{ev}(\theta_a)-1)\phi_a][H_{ev}(\theta_a)\phi_a]^T\} \\ &\quad + O(\|e_a\|_\infty) \end{aligned} \tag{5.32}$$

If $J_\theta(0)$ is small and $J_{\theta\theta}(0) > 0$, then θ_a is approximately optimal, that is, θ_a is close to the parameter setting at which tracking error power is near a local minimum. The conditions for which this approximation is meaningful are essentially the same conditions that we have stated throughout the text, namely:

(C1) small $e_a(t)$

(C2) $|1 - H_{ev}(j\omega,\theta_a)|$ small at those frequencies where $\phi_a(t)$ has significant power

(C3) $\phi_a(t)$ persistently exciting in the "good" frequency range, i.e., so that $\text{Re}\lambda\{\text{avg}[\phi_a H_{ev}(\theta_a)\phi_a^T]\} > 0$

The required smallness of $\|(1-H_{ev}(\theta_a))\phi_a\|$ will appear again in the discrete-time setting (see Chapter 5, Section 4). As we observed in Section 4.4.3, $H_{ev}(s,\theta_a)$ can be expressed as

$$H_{ev}(s,\theta_a) \;=\; H_{\text{ref}}(s) + \tilde{H}_{ev}(s,\theta_a) \tag{5.33}$$

where $\tilde{H}_{ev}(s,\theta_a)$ is the unmatchable part of the closed-loop dynamics. Thus,

$$|1-H_{ev}(j\omega,\theta_a)| \;\le\; |1-H_{\text{ref}}(j\omega)| + |\tilde{H}_{ev}(j\omega,\theta_a)| \tag{5.34}$$

Since $H_{\text{ref}}(0) = 1$ and $\tilde{H}_{ev}(0,\theta_a) = 0$, it follows that $\|(1-H_{ev}(\theta_a))\phi_a\|$ is small if the power in $\phi_a(t)$ is concentrated at low frequencies. Actually, the range of permitted frequencies in $\phi_a(t)$ can be considerably broadened by filtering the regressor and/or error vectors. By use of a more complex adaptive algorithm, as well as filtering the regressor and error signals, it is always possible to obtain [see, e.g., Egardt (1979)]

$$H_{ev}(s,\theta_a) = 1 + \tilde{H}_{ev}(s,\theta_a) \tag{5.35}$$

with $\tilde{H}_{ev}(0,\theta_a) = 0$. The filtered signals are used both within the adaptive control structure as as well as within the parameter algorithm.

To see the effect of filtering as a modification alone, let us now consider the case where the error and regressor signals are filtered *prior* to their use in the parameter algorithm by stable proper filters $W_e(s)$ and $W_\phi(s)$, respectively. Hence, the adaptive error system becomes

$$\begin{aligned} \dot{\theta} &= \epsilon\phi^f e^f \\ \phi^f &= W_\phi \phi \\ e^f &= W_e e \end{aligned} \tag{5.36}$$

where e and ϕ retain their original definition, i.e.,

$$\begin{aligned} e &= e_* - H_{ev}(\phi^T\theta) \\ \phi &= \phi_* - H_{\phi v}(\phi^T\theta) \end{aligned}$$

Thus, the candidate tuned setting θ_a is obtained from

$$\text{avg}\{\phi_a^f e_a^f\} = \text{avg}\{(W_\phi\phi_a)W_e e_a)\} = 0 \tag{5.37}$$

Also, local stability requires small $\|e_a^f\|_\infty$ and

$$\text{Re}\lambda\{\text{avg}[\phi_a^f(W_e H_{ev}(\theta_a)\phi_a)^T]\} = \text{Re}\lambda\{\text{avg}[(W_\phi\phi_a)(W_e H_{ev}(\theta_a)\phi_a)^T]\} > 0 \tag{5.38}$$

Moreover, the mean-square filtered error,

$$J^f(\theta) = \frac{1}{2}\,\text{avg}\{(W_e\bar{e})^2\} \tag{5.39}$$

is near a local minimum for θ_a if

$$J_\theta'(0) = \text{avg}\{(W_e e_a)(W_\phi - W_e H_{ev}(\theta_a))\phi_a\} \tag{5.40}$$

is small. We now see that the error and regressor filters can have somewhat different effects. For example, let the error filter be given by

$$W_e(s) = W_\phi(s)/H_{\text{ref}}(s) \tag{5.41}$$

Using (5.33) we have

$$J_\theta'(0) = -\text{avg}\{(W_e e_a)(W_\phi \tilde{H}_{ev}(\theta_a)/H_{\text{ref}} \phi_a)\} \tag{5.42}$$

and (5.38) becomes

$$\text{Re}\lambda\{\text{avg}[(W_\phi\phi_a)(1 + \tilde{H}_{ev}(\theta_a)/H_{\text{ref}})(W_\phi\phi_a)^T]\} > 0 \tag{5.43}$$

Consequently, it is desirable to select $W_\phi(s)$ so that $|W_\phi(j\omega)\tilde{H}_{ev}(j\omega,\theta_a)/H_{\text{ref}}(j\omega)|$ is small wherever $\phi_a(t)$ has significant power. Of course proper selection of $W_\phi(s)$ requires some knowledge of $\tilde{H}_{ev}(s,\theta_a)$ which in turn means a priori knowledge about the plant to be controlled. Filtering modifications are further discussed with specific examples in Chapter 6.

4.5.2 Linearization and Averaging: Sample Average Signals

In the previous section we considered the case where the tuned signals $e_*(t)$ and $\phi_*(t)$ were almost-periodic functions, and, thus, possess an average value. In this section we relax this assumption and only require that they have a sample-average in the sense defined in Chapter 3 (Section 3.4). That is, let $\{t_k\}$ be an ordered sequence of sample times such that

$$t_{k+1} - t_k \in (0,\infty) \quad , \quad \forall\, k \in \mathbf{Z}_+ \tag{5.44}$$

Corresponding to this sequence, a function $f(t)$ has a *sequence of sample-averages* $\{\bar{f}(k)\}$ defined by

$$\bar{f}(k) = \frac{1}{t_{k+1}-t_k}\int_{t_k}^{t_{k+1}} f(t)dt \quad , \quad \forall\, k \in \mathbf{Z}_+ \tag{5.45}$$

provided that $\bar{f}(\cdot) \in l_\infty$.

Suppose that $e(t,\theta_*)$ and $\phi(t,\theta_*)$ possess the sequence of sample-averages $\bar{e}(k,\theta_*)$ and $\bar{\phi}(k,\theta_*)$, respectively, corresponding to some ordered sequence $\{t_k\}$. Since the tuned signals need not be almost-periodic, the previous analysis in Section 3.1 does not apply. For example, unlike (5.25), in this case it is more different to be precise about a candidate tuned setting

A convenient approach, which we will now discuss, is to pre-select θ_* as a constant tuned setting. Then, on the basis of a linearization in the neighborhood of θ_*, determine conditions on the tuned signals so that during adaptation the system is locally stable, i.e., $\hat{\theta}(t)$ remains in a small neighborhood of θ_*. In fact, we will establish a *region of attraction*, which is not necessarily small, from which $\hat{\theta}(t)$ is guaranteed to exponentially approach a small neighborhood of θ_*. This procedure can also be invoked even if the signals are almost-periodic, in which case θ_* can be obtained from $\text{avg}(e_*\phi_*) = 0$. Hence, the methodology in this section has a more general application. The approach we take is based on linearization, linear averaging of Chapter 3, and the Total Stability Theorem presented in Chapter 1.

We start with the error system (5.1) in the form (5.6), which we repeat completely here for convenience as

$$\dot{\theta} = \epsilon q(t,\bar{z}(t,\theta) + \eta) \tag{5.45a}$$

$$\dot{\eta} = F(\theta)\eta - \epsilon\bar{\psi}(t,\theta)q(t,\bar{z}(t,\theta)+\eta) \tag{5.45b}$$

with

$$q(t,z) = \phi_*e_* - Q_*z + c^TzDz \ , \quad Q_* = \phi_*c^T + e_*D \tag{5.45c}$$

$$F(\theta) = A(\theta_*) - b\theta^TD \quad , \quad G(\theta) = b\theta^T \tag{5.45d}$$

and where

$$\bar{\psi}(t,\theta) = \partial\bar{z}(t,\theta)/\partial\theta \tag{5.46a}$$

with $\bar{z}(t,\theta)$ and $\bar{\psi}(t,\theta)$ defined for each fixed θ as the solutions of

$$\partial\bar{z}/\partial t = F(\theta)\bar{z}+G(\theta)\phi_*(t) \tag{5.46b}$$

$$\partial\bar{\psi}/\partial t = F(\theta)\bar{\psi}+b(\phi_*(t)-D\bar{z})^T \tag{5.46c}$$

In the previous section the signals were assumed to be almost periodic, and hence, initial conditions were subsumed by defining operators from the infinite past to the present, e.g., (5.4). The situation here is different, because the signals are not necessarily almost periodic. Thus, at some initial time (for convenience we choose $t = 0$), the initial states of (5.45) are,

$$\theta(0) = \theta_0 \in \mathbb{R}^p \quad , \quad z(0) = z_0 \in \mathbb{R}^n \tag{5.47}$$

It is convenient to assign for each fixed θ,

$$\bar{z}(0,\theta) = z_0 \tag{5.48}$$

As a result, from the definitions of $\eta(t)$ and $\bar{\psi}(t,\theta)$, we also have,

$$\eta(0) = 0 \in \mathbb{R}^n \quad , \quad \bar{\psi}(0,\theta) = 0 \in \mathbb{R}^{n\times p} \tag{5.49}$$

At this point it is possible to apply the Total Stability Theorem 1.2 with the state $x = \begin{pmatrix}\theta\\ \eta\end{pmatrix}$. The disadvantage to this direct approach is that the conditions of the theorem will hold provided that $\theta(0) = O(\epsilon)$. Because we would like to establish an allowable bound on initial conditions $\theta(0)$ which does not tend to zero with ϵ, we use a two-step approach. In the first step we show that $|\theta(t)| \leq r$ implies that $\eta(t) = O(\epsilon)$. In the second step we use $\eta(t) = O(\epsilon)$ to show that $|\theta(t)| \leq r$, and furthermore, that

$$\limsup_{t\to\infty} |\theta(t)| \leq O(\|e_*\|_\infty)+O(\epsilon) \tag{5.50}$$

In each step we use a modified version of the Total Stability Theorem 1.2 which takes into account exponentially decaying terms. We now state this modified version and then by using the two-step approach outlined above, we will rigorously obtain (5.50).

Theorem 4.3: Total Stability (Time-Varying Bounds)

Consider:

$$\dot{x} = A(t)x+f(t,x) \quad , \quad t \geq 0 \, , \; x(0) = x_0 \tag{5.51}$$

Assume $\exists r > 0$ such that $\forall \, |x| \leq r$:

(A1) $A(t), f(t,x)$ are locally integrable in t (5.52)

(A2) $|f(t,x)| \leq \gamma_0+\gamma_1 e^{-\lambda_1 t}+(\beta_0+\beta_1 e^{-\lambda_1 t})|x|, \forall t \geq 0$ (5.53)

(A3) $\dot{y} = A(t)y$ is u.a.s. i.e., its transition

matrix $F(t,\tau)$ satisfies

$$|F(t,\tau)| \leq m_0 e^{-\lambda_0(t-\tau)}, \forall t \geq \tau \tag{5.54}$$

Under these conditions, if:

$$\lambda = \lambda_0 - \beta_0 m_0 \in (0,\lambda_1) \tag{5.55}$$

$$|x_0| < r/m-\gamma_1/(\lambda_1-\lambda) \tag{5.56}$$

$$m\gamma_0 < \lambda r \quad , \quad m = m_0 e^{m_0\beta_1/\lambda_1} \tag{5.57}$$

then $\exists$ solution $x(t)$ of (5.51) such that $\forall t \geq 0$:

$$|x(t)| \leq m\left(|x_0|+\frac{\gamma_1}{\lambda_1-\lambda}\right)e^{-\lambda t} + \frac{m\gamma_0}{\lambda}(1-e^{-\lambda t}) \leq r \tag{5.58}$$

Remarks: These results differ from those in Theorem 1.2 in that: (1) the bounding functions explicitly contain exponentially decaying terms; (2) existence, but not uniqueness, is established because f is not assumed to be locally Lipschitz in x.

Proof *(Outline)*

We present an outline of the proof. Following the standard arguments such as those found in Bellman (1953), Coppel (1965), and Hale (1969), it can be shown that condition (A1) insures that any solution of (5.51) also satisfies,

$$x(t) = F(t,0)x_0 + \int_0^t F(t,\tau)f(\tau,x(\tau))d\tau \tag{5.59}$$

For those values of $t \geq 0$ for which $|x(t)| \leq r$, say $t \in [0,t_1]$, we have

$$|x(t)| \leq m_0 e^{-\lambda_0 t}|x_0| + \int_0^t m_0 e^{-\lambda_0(t-\tau)}[\gamma_0 + \gamma_1 e^{-\lambda_1\tau} + (\beta_0 + \beta_1 e^{-\lambda_1\tau})|x(\tau)|]d\tau \tag{5.60}$$

Applying the Bellman-Gronwall Lemma results in the bound given by (5.58) which is valid for $t \in [0,t_1]$. Since the bound is uniform in t, it holds for all $t \geq 0$ and the theorem is proved.

□

The first step is to find a bound $|\theta| \leq r$ for which $\bar{z}(t,\theta)$ and $\bar{\psi}(t,\theta)$ are uniformly bounded. It is convenient to express the solutions to (5.46) in their explicit form:

$$\bar{z}(t,\theta) = e^{tF(\theta)}z_0 + \left(\int_0^t e^{(t-\tau)F(\theta)}b\phi_*(\tau)^T d\tau\right)\theta \tag{5.61a}$$

$$\bar{\psi}(t,\theta) = \int_0^t e^{(t-\tau)F(\theta)}b[\phi_*(\tau) - D\bar{z}(\tau,\theta)]^T d\tau \tag{5.61b}$$

Observe that $\bar{\psi}(t,0) = L_0(t)$ is the "L-transformation" matrix of Chapter 3 and (5.13), where here it satisfies,

$$\dot{L}_0 = A(\theta_*)L_0 + b\phi_*^T \ , \quad L_0(0) = 0 \tag{5.62}$$

Using $L_0(t)$ gives the following useful expression for (5.61a):

$$\bar{z}(t,\theta) = e^{tF(\theta)}z_0 + L_0(t)\theta + \tilde{z}(t,\theta) \tag{5.63a}$$

$$\tilde{z}(t,\theta) = -\int_0^t e^{(t-\tau)F(\theta)}b\theta^T DL_0(\tau)\theta d\tau \tag{5.63b}$$

Since $\mathrm{Re}\lambda[A(\theta_*)] < 0, \exists K \geq 1$ and $\alpha > 0$ such that

$$|e^{tA(\theta_*)}| \leq Ke^{-\alpha t} \ , \quad \forall\, t \geq 0 \tag{5.64}$$

As a result we immediately obtain the following result from (5.61)-

(5.62).

Lemma 4.2:

If

$$\mu = \alpha - a^2 r > 0 \tag{5.65}$$

$$a = \max\{|b|, |c|, |D|\} \tag{5.66}$$

then $\forall\, |\theta| \leq r$ and $\forall\, t \geq 0$,

$$\begin{aligned} \left|e^{tF(\theta)}\right| &\leq Ke^{-\mu t} \\ |\tilde{z}(t,\theta)| &\leq (Ka^2\|L_0\|_\infty/\mu)|\theta|^2 := k_0|\theta|^2 \\ |\bar{\psi}(t,\theta)| &\leq (Ka/\mu)[\|\phi_*\|_\infty + a(K|z_0| + \|L_0\|_\infty|\theta| + k_0|\theta|^2)] \\ &:= k_1 + k_2|\theta| + k_3|\theta|^2 \end{aligned} \tag{5.67}$$

□

Thus, if $|\theta| \leq r$ then $\bar{z}(t,\theta)$ and $\bar{\psi}(t,\theta)$ are uniformly bounded. From (5.63) we see that $\bar{z}(t,\theta)$ consists of the L-transformation term $L_0(t)\theta$, the term $e^{tF(\theta)}z_0$ which is exponentially decaying for $|\theta| \leq r$, and the term $\tilde{z}(t,\theta)$ which is $O(|\theta|^2)$. We will see that the effect of the z_0 terms decaying exponentially is significant in determining the region of attraction.

The same results also apply to the time-varying case where θ is replaced with $\theta(t)$ and $|\theta(t)| \leq r$. Note that $\bar{z}(t,\theta(t))$ and $\bar{\psi}(t,\theta(t))$ are obtained by replacing θ with $\theta(t)$ in (5.61) and *not* by replacing θ with $\theta(t)$ in (5.46).

In order to apply Theorem 4.3 and specify the bounds therein, it is necessary to explicitly display the terms in $q(t,\bar{z}(t,\theta)+\eta)$. In an obvious shorthand notation we have,

$$\begin{aligned} q(t,\bar{z}+\eta) &= \phi_* e_* - Q_*(\bar{z}+\eta) + c^T(\bar{z}+\eta)D(\bar{z}+\eta) \\ &= -R(t,\eta,\theta)\theta + g(t,\eta,\theta) \end{aligned} \tag{5.68a}$$

where

$$R(t,\eta,\theta) = R_* - c^T\xi D L_0 - D\xi c^T L_0 \tag{5.68b}$$

$$g(t,\eta,\theta) = \phi_* e_* - Q_*(\xi+\tilde{z}) + c^T\xi D(\xi+\tilde{z})$$
$$+ c^T L_0\theta D(L_0\theta+\tilde{z}) + c^T\tilde{z}D(\xi+L_0\theta+\tilde{z})$$

with $\tilde{z}$ defined in (5.67) and R_* and ξ defined by,

$$R_*(t) := (\phi_* H_{ev}\phi_*^T + e_* H_{\phi v}\phi_*^T)(t) \tag{5.68d}$$

$$\xi(t,\theta,\eta) := e^{tF(\theta)}z_0 + \eta \tag{5.68e}$$

Using Lemma 4.3, if $|\theta| \leq r$ then

$$|\xi| \leq Ke^{-\mu t}|z_0| + |\eta|$$
$$|\tilde{z}| \leq k_0|\theta|^2 \tag{5.69}$$
$$|\overline{\psi}| \leq k_1 + k_2|\theta| + k_3|\theta|^2$$

Now, using the above relations together with Theorem 4.3 we obtain

Lemma 4.3:

Consider system (5.45b) with θ replaced by $\theta(t)$, i.e.,

$$\dot{\eta} = F(\theta(t))\eta - \epsilon\overline{\psi}(t,(\theta(t))q(t,\overline{z}(t,(\theta(t))+\eta)\ ,\ \ \eta(0) = 0 \tag{5.70}$$

If (5.64) holds and

$$|\theta(t)| \leq r < \alpha/(Ka^2) \tag{5.71}$$

then

$$\eta(t) = O(\epsilon) \tag{5.72}$$

Remark: Observe that since $K \geq 1$, the bound in (5.71) insures that (5.65) is satisfied.

Proof

Apply Theorem 4.3 with $x = \eta$ while treating θ as a function of time, and set:

$$A(t) = A(\theta_*) \tag{5.73}$$

$$f(t,\eta) = -b\theta^T D\eta - \epsilon\overline{\psi}(t,\theta)q(t,\overline{z}(t,\theta)+\eta) \tag{5.74}$$

This allows us to take (in the notation of Theorem 4.3):

$$\begin{aligned} \gamma_0 &= O(\epsilon) \quad , \quad \gamma_1 = 0 \\ \beta_0 &= a^2 r + O(\epsilon) \quad , \quad \beta_1 = 0 \\ m_0 &= K \quad , \quad \lambda_0 = \alpha \end{aligned} \tag{5.75}$$

Conditions (5.55)-(5.57) become,

$$\begin{aligned} \lambda &= \alpha - K a^2 r - O(\epsilon) > 0 \\ & 0 < r/K \\ & O(\epsilon) < \lambda r/K \end{aligned} \tag{5.76}$$

Thus, as long as (5.71) holds, (5.58) $\Rightarrow$ $\eta(t) = O(\epsilon)$.

□

We now state the main result of this section.

Theorem 4.4: Local Stability

Consider the adaptive system (5.6). Assume that:

$$\text{(A1)} \quad |e^{tA(\theta_*)}| \le K e^{-\alpha t} \quad , \quad \forall t \ge 0 \tag{5.77}$$

(A2) $\dot{\theta} = -\epsilon R_*(t)\theta$ with $R_*(t)$from (5.68d) is u.a.s.

$\forall$ small $\epsilon > 0$, with transition matrix

$$|F(t,\tau)| \le M e^{-\epsilon\sigma(t-\tau)} \quad , \quad \forall t \ge \tau \ge 0 \tag{5.78}$$

Under these conditions, for any $|z_0| < \infty$, if:

$$r < \alpha/(K a^2) \tag{5.79a}$$

$$|\theta_0| < r/M \tag{5.79b}$$

$$M \|\phi_* e_*\|_\infty < (\sigma - M\rho(r)) r \tag{5.79c}$$

where

$$\rho(r) = r[k_0 \|Q_*\|_\infty + a^2(\|L_0\|_\infty + k_0 r)^2] \tag{5.80}$$

then $\exists$ small $\epsilon > 0$ such that

$$\sup_t |\eta(t)| = O(\epsilon) \tag{5.81}$$

$$\sup_t |\theta(t)| \leq r \tag{5.82}$$

$$\lim_{t\to\infty} \sup |\theta(t)| \leq \frac{M}{\sigma - M\rho(r)} \|\phi_* e_*\|_\infty + O(\epsilon) \tag{5.83}$$

Moreover, the convergence of $\theta(t)$ to the set defined by (5.83) is exponential at the rate

$$\lambda_\theta = \epsilon[\sigma - M\rho(r)] \tag{5.84}$$

Proof

We start with the *temporary* assumption that $|\theta(t)| \leq r$. From Lemma 4.3 we have that $\eta(t) = O(\epsilon)$. Now apply Theorem 4.3 with $x = \theta$ while treating η as a function of time, and set:

$$A(t) = -\epsilon R_*(t) \tag{5.85}$$

$$f(t,\theta) = \epsilon[(R(t,\eta,\theta) - R_*(t))\theta + g(t,\eta,\theta)] \tag{5.86}$$

Using the expressions (5.68)-(5.69) together with $\eta(t) = O(\epsilon)$, we can take:

$$\gamma_0 = \epsilon\|\phi_* e_*\|_\infty + O(\epsilon^2)$$

$$\gamma_1 = |z_0|O(\epsilon) \quad , \quad \lambda_1 = \alpha - Ka^2 r$$

$$\beta_0 = O(\epsilon^2) + \epsilon\rho(r)$$

$$\beta_1 = |z_0|O(\epsilon) + r|z_0|O(\epsilon)$$

$$m_0 = M \quad , \quad \lambda_0 = \epsilon\sigma \quad , \quad m = Me^{O(\epsilon)}$$

To show that the quantity $\rho(r)$ above is actually that stated in (5.80), observe that from the definition (5.53),

$$\begin{aligned}\rho(r)|\theta| &\geq \sup_{|\theta|\leq r} \lim_{t\to\infty} \sup |g(t,0,\theta) - g(t,0,0)| \\ &= \sup_{|\theta|\leq r} \sup_{t\geq 0} |-Q_*\tilde{z} + c^T(L_0\theta + \tilde{z})D(L_0\theta + \tilde{z})|\end{aligned}$$

where $\tilde{z}$ is bounded as in (5.69). The expression for $\rho(r)$ then follows. Conditions (5.55)-(5.57) then become:

$$\lambda = \epsilon\sigma - \epsilon M[O(\epsilon)+\rho(r)] \in (0,\alpha-Ka^2r)$$

$$|\theta_0| < r/m - |z_0|O(\epsilon)/(\alpha-Ka^2r-O(\epsilon))$$

$$\epsilon m[\|\phi_* e_*\|_\infty + O(\epsilon)] < r$$

It is clear that these relations can be satisfied for small $\epsilon > 0$ provided (5.79) holds. Finally, but still under the temporary assumption that $|\theta(t)| \leq r$, (5.81)-(5.83) follows from (5.58).

The temporary assumption can be removed by following standard arguments which establish a contradiction, e.g., Bellman (1953). Note first that (5.79b) implies $|\theta_0| < r$ because $M \geq 1$. Hence, under the assumptions of the theorem, it follows that there exists $t_1 > 0$ such that (5.81) holds for $t \in [0,t_1]$. This implies that (5.58) holds with $x(t)$ replaced by $\theta(t)$ together with the appropriate replacement of the various constants in (5.58). Suppose by way of contradiction that $|\theta(t)| < r$ for some $t > 0$. It follows then that we can take t_1 as the first time that $|\theta(t_1)| = r$. Note that $t_1 > 0$ because $|\theta_0| < r$. Now, using conditions (5.79) and (5.58) with $x(t)$ replaced by $\theta(t)$, it follows that for some small $\epsilon > 0$, $|\theta(t_1)| < r$. But this contradicts the supposition that $|\theta(t_1)| = r$, and hence, $|\theta(t)| < r$, $\forall t \in [0,t_1]$. Since the bound is uniform the argument can be repeated on an interval starting with t_1 until (5.81)-(5.83) are established $\forall t \geq 0$, which completes the proof.

□

Discussion

The conditions and assumptions of Theorem 4.4 restrict the allowable signals and initial conditions of the adaptive system. Firstly, we require $\dot{\theta} = -\epsilon R_*(t)\theta$ to be u.a.s. for all small $\epsilon > 0$, where from (5.68d)

$$R_*(t) = (\phi_* H_{ev} \phi_*^T + e_* H_{\phi v} \phi_*^T)(t) \tag{5.87}$$

Recall from Chapter 3 (Section 3.4) that the above u.a.s. property holds for some ordered sequence of sample times $\{t_k\}$ if the sequence of sample-averages,

$$\bar{R}_*(k) = \frac{1}{t_{k+1}-t_k} \int_{t_k}^{t_{k+1}} R_*(t)dt \ , \quad \forall k \in \mathbf{Z}_+ \tag{5.88}$$

is such that $\exists$ constant matrix $P = P^T > 0$ which satisfies

$$P\bar{R}_*(k) + \bar{R}_*(k)^T P \geq I \ , \quad \forall k \in \mathbf{Z}_+ \tag{5.89}$$

This statement of the signal dependent positivity condition imposes restrictions not only on the tuned regressor ϕ_*, but also on the tuned error e_*. Additional restrictions on ϕ_* and e_* arise from (5.79), and moreover, σ and M are in turn dependent on the properties of $\bar{R}_*(k)$ as restricted by (5.89).

A very interesting aspect of Theorem 4.4 is that as long as (5.79) holds, (5.81)-(5.83) follow for any initial condition $|z_0| < \infty$ and some small $\epsilon > 0$. Also, provided $\|e_*\|_\infty$ is small, θ_0 is limited to the constant parameter stability set. The analysis as such is thus limited to adaptive control of systems which are initially stable. Nonetheless, conditions (5.79) defines a not necessarily small region of attraction for (5.83).

The allowable range of θ_0 is not only restricted to be in the constant parameter stability set but it is further limited by the effect of the nonlinear terms. These are reflected in the size of $\rho(r)$ and condition (5.79c), where even if $\|\phi_* e_*\|_\infty$ is small, it is still necessary that $\sigma - M\rho(r) > 0$. Suppose, for example, that there is an r satisfying (5.79) such that $r - \alpha/Ka^2$ is small. Since $k_0 \sim 1/(r-\alpha/a^2)$, it follows that $\rho(r)$ becomes significant. If $r << \alpha/(Ka^2)$, then $\rho(r)$ will have considerably less affect since $\rho(r) \sim r$ which is now small. In this latter case $|\theta_0|$, the initial parametrization, must be close to the tuned setting θ_*.

Theorem 4.4 also asserts that the rate of convergence is exponential and proportional to ϵ. Thus, if $|z_0|$ or $|\theta_0|$ are large, a small required ϵ will provide slow but exponential convergence.

Another aspect of convergence is the uniformity of convergence with respect to all the parameters. More specifically, consider the homogeneous linear system $\dot{\theta} = -\epsilon R_*(t)\theta$. From Chapter 3, if (5.89) holds, then $\forall$ small $\epsilon > 0$, we have the sequence of contractions

$$\theta(t_{k+1})^T P \theta(t_{k+1}) < \theta(t_k)^T P \theta(t_k) \quad , \quad \forall k \in \mathbf{Z}_+ \tag{5.90}$$

Thus, the condition number of P (the ratio of max to min eigenvalues) provides information as to the uniformity of convergence over the parameters or combinations thereof. If all the parameters are equally important to performance, then the ideal case is to have $P = I$. Thus,

$$\bar{R}_*(k) + \bar{R}_*(k)^T \geq I \quad , \quad \forall k \in \mathbf{Z}_+ \tag{5.91}$$

insures that all parameters converge at the same rate. This can be taken as a *rule of design*. In the general case one may choose as a norm on the vector space

$$|x| = (x^T P x)^{1/2} \tag{5.92}$$

where P satisfies (5.89). Hence, all the bounds in Theorem 4.4 are generated from this norm, consequently the expressions in Theorem 4.4 are quite interconnected. In the special case when $P = I$, some of the complexity dissolves because this particular choice is really hiding the fact that the designer has available a good deal of *a priori* system information. For example, the constant parameter stability set,

$$\Theta_s = \{\theta \in \mathbb{R}^p : \mathrm{Re}\lambda[F(\theta)] < 0\} \tag{5.93}$$

although quite complicated to describe, may in some instances be well approximated by the convex subset,

$$\Theta_c = \{\theta \in \Theta_s : \theta^T P_c \theta \leq 1\} \tag{5.94}$$

Using the norm (5.92) with P from (5.89), the subset in Θ_s of permissable initial conditions is from (5.79),

$$\Theta_0 = \{\theta \in \Theta_s : \theta^T P \theta < (r_{max}/M)^2\} \tag{5.95}$$

where r_{max} is the largest value of r satisfying (5.79). To make the most use of allowable initial conditions which are in Θ_c requires that P_c and $(M/r_{max})^2 P$ be nearly identical. Again, the choice of $P = I$ simplifies the algebra, but belies the fact that considerable information is available about the true system behavior. This is not at all unreasonable, since such data is often available. Theorem 4.4 thus provides analytic justification for these intuitive notions.

The results of Theorem 4.4 can also be interpreted in terms of the actual regressor and error signals $\phi(t)$ and $e(t)$, respectively. Recall from (1.5)-(1.7) that

$$\begin{aligned} \phi - \phi_* &= -Dz = -H_{\phi v}(\theta^T \phi) \\ e - e_* &= -c^T z = -H_{ev}(\theta^T \phi) \end{aligned} \tag{5.96}$$

In terms of the operators H_{ev} and $H_{\phi v}$, that part of condition (5.79) which requires $r < \alpha/(Ka^2)$ is equivalent to

$$\begin{aligned} r &< 1/\gamma \\ \gamma &= \max\{\gamma_\infty(H_{ev}), \gamma_\infty(H_{\phi v})\} \end{aligned} \tag{5.97}$$

Expressing (5.83) as

$$\begin{aligned} \limsup_t |\theta(t)| &\leq N \|\phi_*\|_\infty \|e_*\|_\infty + O(\epsilon) \\ N &= M/(\sigma - \rho(r)M) \end{aligned} \tag{5.98}$$

gives

$$\begin{aligned} \limsup_t |\phi(t) - \phi_*(t)| &\leq E \|e_*\|_\infty \\ \limsup_t |e(t)| &\leq \|e_*\|_\infty (1+E) \end{aligned} \tag{5.99}$$

where

$$E = \frac{\gamma N \|\phi_*\|_\infty^2}{1 - \gamma N \|\phi_*\|_\infty \|e_*\|_\infty} \tag{5.100}$$

The quantity E gives a measure of the error in performance which could be expected from the action of the adaptive system. In other words, $\|e_*\|_\infty$ needs to be sufficiently small so that the worst case performance deviation $(1+E)\|e_*\|_\infty$, is still within specifications.

In the conditions of Theorem 4.4, a small $\epsilon > 0$ is required. The theorem as presented is qualitative in that *some* range of small $\epsilon > 0$ will suffice, but a specific bound on ϵ is not given although it could be extracted from the proof. Note that this bound depends on the allowable bound on ϵ for which $\dot{\theta} = -\epsilon R_*(t)\theta$ is u.a.s. The inequalities involved are stated in Chapter 3 in Theorem 3.1. The reader should be aware, however, that it may not be possible to satisfy all these inequalities for every choice of θ_*. For example, some choices may make $\|e_*\|_\infty$ too large, even though $\dot{\eta} = A(\theta_*)\eta$ and $\dot{\theta} = -\epsilon R_*(t)\theta$ are both u.a.s. Note that the choice of θ_* can be any setting determined by the user from any criterion, e.g., $\text{avg}(\phi_* e_*) = 0$ is a possibility. Of course this means $e_*(t)$ and $\phi_*(t)$ are almost-periodic; hence, the nonlinear averaging analysis of Section 4.5.1 applies. In this regard Theorem 4.4 provides valuable information because it establishes a bound on the limit set of the parameter error (5.83). Recall that Theorem 4.2 provides for the *existence* of an almost periodic solution which is in an $O(\epsilon)$-neighborhood of $(\theta_a,0)$ where $\text{avg}\{\phi(\cdot,\theta_a)e(\cdot,\theta_a)\} = 0$. Theorem 4.4 then provides the region of attraction to this solution.

In summary: we have analyzed the adaptive system in a not necessarily small neighborhood of either some tuned setting θ_* selected by the user or some tuned setting θ_a selected by the algorithm. We have shown that under slow adaptation, if: (1) the adaptable parameters are initially in a subset of the constant parameter stability set, (2) a frequency dependent positivity condition is satisfied, and (3) the tuned error is sufficiently small, then the parameter will converge slowly, but exponentially, to a neighborhood of θ_*, the size of which depends on the adaptation gain ϵ and the peak tuned error $\|e_*\|_\infty$. What we have *not* established are conditions under which the system will arrive at one of these neighborhoods in the constant parameter stability set from which it will converge to an $O(\|e_*\|_\infty)$-neighborhood which contains a tuned setting. This is another aspect to determining the *region of attraction* of the adaptive system and it is precisely here that our analysis is conservative. Theorem 4.4 shows that $\theta(0)$ need *not* be of order ϵ; thus, a region of attraction is, to some extent, provided. However, as mentioned before, this region is a subset of the constant parameter stability set -- which is certainly to be an expected restriction of this analysis of slow adaptation. That is, initial parametrizations must stabilize the plant. One may speculate that rapid adaptation, at least initially, is what is called for to adaptively stabilize an unstable plant, then to be followed by slow adaptation. In a recursive least squares algorithm with a forgetting factor, this type of gain variation is usually what takes place.

4.6 NOTES AND REFERENCES

The development of the error models presented in Sections 4.1 and 4.2 follows from the standard works in this area, e.g., Egardt (1979), Landau (1976), Monopoli (1974), Narendra, Lin, and Valavani (1980). These error model forms presume an exact matching condition is satisfied, which is not the case in our analysis. Hence, in Section 4.4 we form the error model for a few simpler cases where exact matching is impossible. This treatment follows that presented in Kosut and Friedlander (1982, 1985) and Kosut and Johnson (1984). The global stability analysis which uses the Popov-Kalman-Yakubovic Lemma can be traced back to the hyperstability treatment in Monopoli (1974), Landau (1976), and Narendra, Lin, and Valavani (1980). These analyses consider more complicated adaptive algorithms than posed here, but the principal ideas are essentially the same. The form of the Global Stability Theorem (Theorem 4.1) is taken from Kosut and Friedlander (1982, 1985) where the proof is based on input-output passivity theory, and hence the theorem statement remains valid for linear systems with irrational transfer functions. The instability examples in Section 4.4.4 are from Ioannou and Kokotovic (1984), which contains other examples of instability mechanisms as well as an algorithm modification to reduce their impact.

The original work on applying averaging theory to the linearized adaptive system is in Riedle and Kokotovic (1985a). Extensions to the nonlinear case, along with the introduction of the integral manifold are in Riedle and Kokotovic (1985b) and in Praly (1985a) for the discrete-time case. The results in Section 4.5.1 follow both these references, particularly in regard to the meaning of $\text{avg}(\phi_* e_*) = 0$. The frozen parameter system was introduced in Astrom (1983, 1984). Related work on averaging for stochastic systems can be found in Ljung and Soderstrom (1983). Some basic references for the classical nonlinear averaging analysis are Hale (1969, 1980) and Volosov (1963). Recent approaches to nonlinear averaging, specifically tailored for adaptive systems, appear in Fu, Bodson, and Sastry (1985), and in Bodson, et al. (1985). Closely related to averaging theory is critical systems theory (Hale, 1980) which is used in the discrete-time periodic case by Praly (1985b).

The analysis in 4.5.2 is partly new, but is based on the earlier work on local stability and linearization in Kosut and Johnson (1984) and Kosut and Anderson (1985), where the set-up is in terms of input-output operator properties. Similar approaches which also provide an estimate of the region of attraction can be found in Anderson and Johnstone (1983) and Ioannou and Kokotovic (1983).

Chapter 5

STABILITY OF DISCRETE-TIME ADAPTIVE SYSTEMS

5.1 PARAMETER ESTIMATION

5.1.1 Introduction

In this section we shall examine first one of the simplest adaptive systems problems -- that of equation error parameter estimation. Then we shall move to the slightly more complex output error identification. While these results are of interest in themselves, they are also very useful in the sense of providing background for the adaptive control problems which are the main theme of the chapter.

5.1.2 Equation Error Parameter Estimation

We suppose there is given a plant

$$A(q^{-1})y(k) = q^{-d}B(q^{-1})u(k) \tag{1.1}$$

where $u(\cdot), y(\cdot)$ are the input and output, $A(0) = 1$, $B(0) \neq 0$, and q^{-1} is the unit delay operator. The degrees of $A(\cdot)$, $B(\cdot)$ are assumed known as α, β, respectively, and the plant is assumed stable; otherwise $A(\cdot)$ and $B(\cdot)$ are unknown. The input is assumed bounded; because the plant is stable, the output also will be bounded. The adaptive equation error identifier, probably well-known to most readers, is constructed in the following way. We start with the definitions:

$$A(q^{-1}) = 1 + \sum_{l=1}^{\alpha} a_l q^{-l}, \quad B(q^{-1}) = \sum_{l=0}^{\beta} b_l q^{-l} \tag{1.2}$$

$$\theta_0 = [b_o \cdots b_\beta \;\; -a_1 \cdots -a_\alpha]^T \tag{1.3}$$

$$\hat{\theta}(k) = [\hat{b}_0(k) \cdots \hat{b}_\beta(k) \;\; -\hat{a}_1(k) \cdots -\hat{a}_\alpha(k)]^T \tag{1.4}$$

Here, θ_0 is the true, initially unknown parameter vector, and $\hat{\theta}(k)$ an estimate of it, assumed available at time k just before measurement of $y(k)$. Define also the regression vector:

$$\phi(k) = [u(k-d) \cdots u(k-d-\beta)\; y(k-1) \cdots y(k-\alpha)]^T \tag{1.5}$$

The plant equation may be rewritten as

$$y(k) = \phi^T(k)\theta_0 \tag{1.6}$$

Beginning with an arbitrary initial estimate of θ_0, call it $\hat{\theta}(0)$, we use the error between $y(k)$ and what we estimate as $y(k)$ while using our current estimate $\hat{\theta}(k)$ of θ_0, i.e., $y(k)-\phi^T(k)\hat{\theta}(k)$, to update the estimate of θ_0, for example, by

$$\hat{\theta}(k+1) = \hat{\theta}(k)+\mu\phi(k)[y(k)-\phi^T(k)\hat{\theta}(k)] \tag{1.7}$$

Here μ is a positive constant, and Equation (1.7) is an example of a gradient algorithm. One can also (and preferably) use a normalized gradient algorithm:

$$\hat{\theta}(k+1) = \hat{\theta}(k) + \frac{\mu\phi(k)}{1+\nu\phi^T(k)\phi(k)}[y(k)-\phi^T(k)\hat{\theta}(k)] \tag{1.8}$$

Here ν is a positive constant, with $\nu \geq \mu$.

If a least squares algorithm is used, then

$$\hat{\theta}(k+1) = \hat{\theta}(k) + \frac{P(k)\phi(k)}{1+\phi^T(k)P(k)\phi(k)}[y(k)-\phi^T(k)\hat{\theta}(k)] \tag{1.9a}$$

$$P(k+1) = P(k) - \frac{P(k)\phi(k)\phi^T(k)P(k)}{1+\phi^T(k)P(k)\phi(k)} \tag{1.9b}$$

and $P(0)$ is generally a large positive definite matrix.

If least squares with exponential forgetting is used,

$$\hat{\theta}(k+1) = \hat{\theta}(k) + \frac{\rho P(k)\phi(k)}{\lambda + \rho\phi^T(k)P(k)\phi(k)}\left[y(k) - \phi^T(k)\hat{\theta}(k)\right] \tag{1.10a}$$

$$P(k+1) = \lambda^{-1}\left[I - \frac{\rho P(k)\phi(k)\phi^T(k)}{\lambda + \rho\phi^T(k)P(k)\phi(k)}\right]P(k) \tag{1.10b}$$

for $\rho \in (0,1)$, $\lambda \in (0,1)$, and least squares with covariance resetting is obtained by taking (1.10) with $\lambda = 1$, but using (1.10a) only for $k \neq mK$, $m = 0,1,2,\ldots$, and K fixed. For $k = mK$, set

$$P(mK) = \eta_m I \tag{1.11}$$

for $0 < \eta_{\min} \leq \eta_m \leq \eta_{\max} < \infty$. Figure 5.1 illustrates the general approach.

The theoretical analysis of these various algorithms is similar. Their practical performance can vary substantially. We shall analyze only the normalized gradient algorithm in any detail, as there is less obscuring algebra. But let us note some of the key practical issues distinguishing the various algorithms.

- The unnormalized gradient method is only practical when signal levels are a priori known.
- Least squares algorithms are faster than gradient algorithms, but more complicated to implement.
- Ordinary least squares generally lacks the ability to track plant parameter variations.
- Least squares with exponential forgetting will be unsatisfactory if there are intervals of little or no plant excitation.

Let us now return to (1.8). Though one *implements* (1.8), one *analyzes* a related equation. Set

$$\theta(k) = \theta_0 - \hat{\theta}(k) \tag{1.12}$$

Recalling (1.6), we obtain from (1.8)

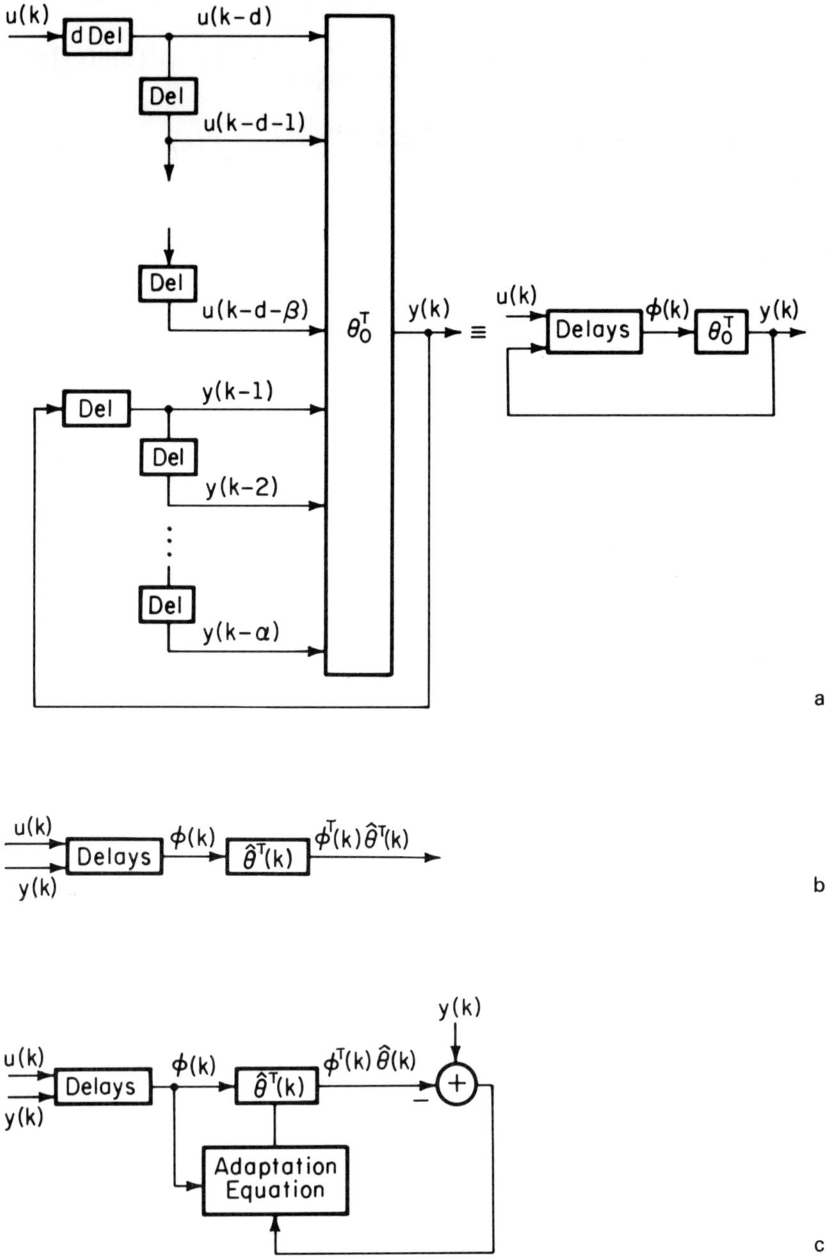

Fig. 5.1 Equation error (a) Plant representation (b) model representation, (c) update law.

$$\theta(k+1) = \theta(k) - \frac{\mu\phi(k)\phi^T(k)}{1+\nu\phi^T(k)\phi(k)}\,\theta(k) \tag{1.13a}$$

$$= \left[I - \frac{\mu\phi(k)\phi^T(k)}{1+\nu\phi^T(k)\phi(k)}\right]\theta(k) \tag{1.13b}$$

The matrix multiplying $\theta(k)$ in (1.13b) has all eigenvalues at 1, except for one eigenvalue which is

$$1 - \frac{\mu\phi^T\phi}{1+\nu\phi^T\phi}$$

By taking $\nu \geq \mu$, we ensure that this eigenvalue lies in (0,1).

As we know from the material of Chapter 2 (Section 2.6), (1.13b) implies:

- $\phi^T(k)\theta(k) \to 0$ as $k \to \infty$, i.e., the error between the true plant output $y(k) = \phi^T(k)\theta_0$ and the output of the estimate of the plant, $\phi^T(k)\hat{\theta}(k)$, approaches zero -- irrespective of $\phi(k)$.
- If for some positive α_1, integer S, and every integer j there holds

$$\sum_{k=j}^{j+S} \phi(k)\phi^T(k) > \alpha_1 I \tag{1.14}$$

then $\|\theta(k)\| \to 0$ at an exponentially fast rate.

The question arises as to when (1.14) holds. Analogously to the continuous time result of Theorem 2.7 in Chapter 2 it can be established that, given the assumption that $A(\cdot)$ and $B(\cdot)$ are coprime (and this is crucial), then (1.14) holds if and only if for some $\alpha_2 > 0$ and all j

$$\sum_{j}^{j+S-\alpha+1} [u(k)\; u(k-1)\; \cdots\; u(k-\alpha-\beta+1)]^T \tag{1.15}$$

$$[u(k)\; u(k-1)\; \cdots\; u(k-\alpha-\beta+1)] > \alpha_2 I$$

The proof of this result will not be given here. A proof of a very similar result, relevant to the study of adaptive control, will be given in a subsequent section of this chapter. (See Lemma 5.3 of Section 5.4.1.)

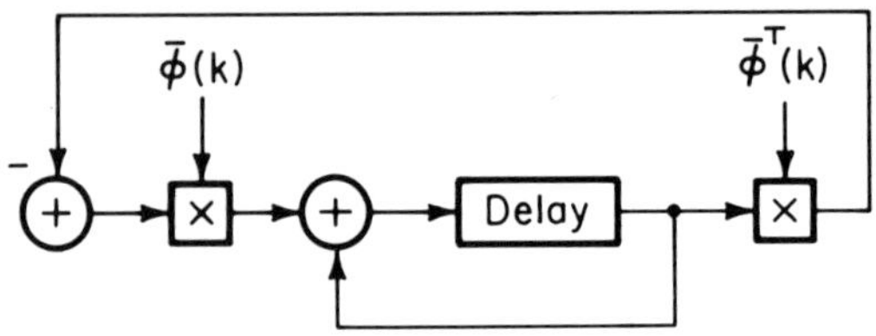

Fig. 5.2 Causal form of update law.

Several other points should be made concerning (1.13b). First, not only does it look linear, but it is linear, since $\phi(k)$ is predetermined by the data (and θ_0), and is independent of $\theta(k)$. Second, Figure 5.2 provides a diagrammatic representation of the equation, with

$$\bar{\phi}(k) \quad := \quad \frac{(\mu)^{1/2}\phi(k)}{(1+\nu\phi^T(k)\phi(k))^{1/2}} \tag{1.16}$$

Third, with

$$\bar{\bar{\phi}}(k) \quad := \quad \frac{(\mu)^{1/2}\phi(k)}{(1+(\nu-\mu)\phi^T(k)\phi(k))^{1/2}} \tag{1.17}$$

one can obtain from (1.13b) the equation

$$\theta(k+1) \;=\; \theta(k) - \bar{\bar{\phi}}(k)\bar{\bar{\phi}}^T(k)\theta(k+1) \tag{1.18}$$

which has the diagrammatic representation of Figure 5.3. Note that when $\nu = \mu$, $\bar{\bar{\phi}}$ is precisely $(\mu)^{1/2}\,\phi$ and (1.18) becomes

$$\theta(k+1) \;=\; \theta(k) - \mu\phi(k)[\phi^T(k)\theta(k+1)] \tag{1.19}$$

The quantity $\phi^T(k)\theta(k+1)$ is an *a posteriori* error, i.e., the error between $y(k)$ and the best model available of the plant *after* $y(k)$ has been used to update the plant parameter estimate $\hat{\theta}(k)$ to $\hat{\theta}(k+1)$.

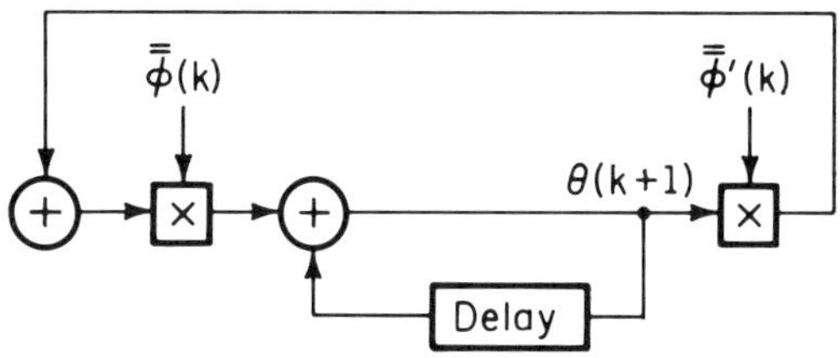

Fig. 5.3 Noncausal form of update law.

Equation (1.19) ties nicely to the ideas of Chapter 1. There, we introduced the equations of the form

$$e = e_* - H_{ev}v \tag{1.20a}$$

$$\phi = \phi_* - H_{\phi v}v \tag{1.20b}$$

$$v = \phi^T\theta \tag{1.20c}$$

$$\dot{\theta} = \epsilon(\Gamma\phi)e \tag{1.20d}$$

The best parallel to these, and in particular (1.20d), is provided by (1.19), which is like a *backward Euler* approximation of (1.20d), when we make the connections $-e(t) = v(t) \leftrightarrow \phi^T(k)\theta(k+1)$, $H_{ev} \leftrightarrow 1$, $e_* \leftrightarrow 0$, $\phi_* = \phi$, $H_{\phi v} = 0$, and $\theta(t) \leftrightarrow \theta(k+1)$ or $\theta(k)$, as appropriate. The parallel between (1.20d) and (1.13b) is nowhere near as close as that between (1.20d) and (1.19).

5.1.3 Output Error Parameter Estimation

In output error identification one of the key ideas is to vary the model depicted in Figure 5.1b in connection with equation error identification so that the plant output $y(k)$ is no longer used as an input to the model; rather, the model output is used. This is of course analogous with the continuous-time case treated in the previous chapter. As before, (1.1) is the plant. We assume available also a known polynomial in q^{-1},

$$C(q^{-1}) = 1 + \sum_{i=1}^{\alpha} c_i q^{-i} \tag{1.21}$$

such that the transfer function

$$H(q^{-1}) = \frac{1+\sum_{i=1}^{\alpha} c_i q^{-i}}{1+\sum_{i=1}^{\alpha} a_i q^{-i}} = A^{-1}(q^{-1})C(q^{-1}) \tag{1.22}$$

is discrete strictly positive real. Thus, in addition to the stability requirement there holds

$$\mathrm{Re}H(e^{j\omega}) > 0 \quad \forall\ \omega \in [0,2\pi) \tag{1.23}$$

We suppose as before that $\theta_0, \hat{\theta}(k)$ denote respectively the true parameter vector and the estimated parameter vector at time k, just before measurement of $y(k)$. The estimate $\hat{\theta}(k+1)$ becomes available just after measurement of $y(k)$. Suppose the model is defined by

$$z(k) = -\sum_{i=1}^{\alpha} \hat{a}_i(k+1)z(k-i) + \sum_{i=0}^{\beta} \hat{b}_i(k+1)u(k-d-i) \tag{1.24}$$

We term $z(k)$ the *a posteriori* model output at time k. We also denote by $\hat{y}(k)$ the *a priori* model output at time k. This is given by

$$\hat{y}(k) = \sum_{i=1}^{\alpha} \hat{a}_i(k)z(k-i) + \sum_{i=0}^{\beta} \hat{b}_i(k)u(k-d-i) \tag{1.25}$$

Now the model output deduced with *a posteriori* parameter values is used instead of $y(k)$ in Figure 5.1b. Hence the regression vector $\phi(k)$ is given now by

$$\phi(k) = [u(k-d) \cdots u(k-d-\beta)\ z(k-1) \cdots z(k-\alpha)] \tag{1.26}$$

(compare (1.5)), and we can rewrite (1.24) and (1.25) as

$$z(k) = \phi^T(k)\hat{\theta}(k+1) \tag{1.27}$$

$$\hat{y}(k) = \phi^T(k)\hat{\theta}(k) \tag{1.28}$$

The update mechanism for $\hat{\theta}(k)$ does not rely simply on either $y(k)-\hat{y}(k)$ or $y(k)-z(k)$. Rather, we define a ***generalized a posteriori*** error between the plant and model by

$$v(k) = C(q^{-1})[y(k)-z(k)] \tag{1.29}$$

Formally, the parameter update equation we shall adopt is

$$\hat{\theta}(k+1) = \hat{\theta}(k) + \mu\phi(k)C(q^{-1})[y(k)-z(k)] \tag{1.30}$$

We have used the word formally because (1.30) in the form given is implicit in $\hat{\theta}(k+1)$ [since $z(k)$ involves $\hat{\theta}(k+1)$]. Equation (1.30) proves useful in later theoretical *analysis* but for *implementation* purposes we must recognize that

$$\begin{aligned} C(q^{-1})[y(k)-z(k)] &= C(q^{-1})y(k)-z(k) - \sum_{i=1}^{\alpha} c_i z(k-i) \\ &= C(q^{-1})y(k)-\phi^T(k)\hat{\theta}(k+1) - \sum_{i=1}^{\alpha} c_i z(k-i) \\ &= C(q^{-1})y(k)-\phi^T(k)\hat{\theta}(k)-\mu\phi^T(k)\phi(k) \\ &\quad \times C(q^{-1})[y(k)-z(k)] - \sum_{i=1}^{\alpha} c_i z(k-i) \end{aligned}$$

i.e.,

$$\begin{aligned} C(q^{-1})[y(k)-z(k)] &= \frac{1}{1+\mu\phi^T(k)\phi(k)} \\ &\quad \times [C(q^{-1})y(k)-\hat{y}(k)- \sum_{i=1}^{\alpha} c_i z(k-i)] \end{aligned}$$

We may regard

$$v_0(k) = C(q^{-1})y(k)-\hat{y}(k) - \sum_{i=1}^{\alpha} c_i z(k-i) \tag{1.31}$$

as a generalized *a priori* error; it is available before $\hat{\theta}(k)$ is updated, and evidently related to the a posteriori error by

$$v(k) = \frac{v_0(k)}{1+\mu\phi^T(k)\phi(k)} \tag{1.32}$$

The practical way to *implement* (1.30) is then to use the a priori error:

$$\hat{\theta}(k+1) = \hat{\theta}(k) + \frac{\mu\phi(k)v_0(k)}{1+\mu\phi^T(k)\phi(k)} \tag{1.33}$$

On the other hand, the *analysis* of this algorithm is achieved most easily by working with a parameter error equation which follows from (1.30). First, with

$$\theta(k) = \theta_0 - \hat{\theta}(k)$$

as before, we have

$$\theta(k+1) = \theta(k) - \mu\phi(k)C(q^{-1})[y(k)-z(k)] \tag{1.34}$$

Now, in obvious notation,

$$\begin{aligned} y(k)-z(k) &= [I-A(q^{-1})]y(k)+B(q^{-1})u(k-d)-[I-\hat{A}(k+1,q^{-1})]z(k) \\ &\quad - \hat{B}(k+1,q^{-1})u(k-d) \\ &= [I-A(q^{-1})][y(k)-z(k)]+[\hat{A}(k+1,q^{-1})-A(q^{-1})]z(k) \\ &\quad + [B(q^{-1})-\hat{B}(k+1,q^{-1})]u(k-d) \\ &= [I-A(q^{-1})][y(k)-z(k)] + \phi^T(k)\theta(k+1) \end{aligned}$$

or

$$y(k)-z(k) = A^{-1}(q^{-1})\phi^T(k)\theta(k+1) \tag{1.35}$$

so that (1.34) becomes

$$\theta(k+1) = \theta(k)-\mu\phi(k)H(q^{-1})[\phi^T(k)\theta(k+1)] \tag{1.36}$$

with $H(q^{-1}) = A^{-1}(q^{-1})C(q^{-1})$. This is the equation used in analyzing the adaptive algorithm.

Let us now draw parallels with the continuous-time equation set (1.20). Of course, (1.35) parallels a backward Euler approximation to (1.20d), and $\phi(t) \leftrightarrow \phi(k)$, $v(t) \leftrightarrow \phi^T(k)\theta(k+1)$, $H_{ev} \leftrightarrow H(q^{-1})$, $e(t) \leftrightarrow -H(q^{-1})[\phi^T(k)\theta(k+1)]$, $e_* \leftrightarrow 0$. Also, $\phi_*(t) \leftrightarrow [u(k-d) \cdots u(k-d-\beta) \cdots y(k-\alpha)]^T$, this being the regression vector obtained when there is correct tuning, i.e.,

$\hat{\theta}(k) \equiv \theta_0$. Also, comparing (1.20b) and (1.35) we see that

$$H_{\phi\nu}(q^{-1}) = [0 \cdots 0 \;\; q^{-1}A^{-1}(q^{-1}) \cdots q^{-\alpha}A^{-1}(q^{-1})]^T \qquad (1.37)$$

Note that $H_{\phi\nu}(q^{-1})$ is a stable transfer function matrix.

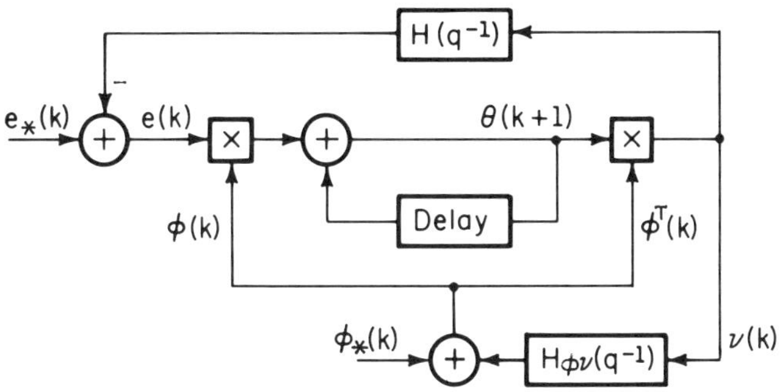

Fig. 5.4 Output error identification.

We reiterate that (1.36) is no longer linear, not of course because of the presence of $H(q^{-1})$, but because $\phi(k)$ depends on $\hat{\theta}(k) = \theta_0 - \theta(k)$.

Figure 5.4 illustrates (1.36). The symbol $\phi_*(k)$ is used for the tuned regression vector, viz.,

$$\phi_*(k) = [u(k-d) \cdots u(k-d-\beta) \;\; y(k-1) \cdots y(k-\alpha)]^T \qquad (1.38)$$

Parenthetically, we remark that the concept of a tuned system for adaptive identification is the system obtained when the adjustable parameters of the adjustable model are fixed at some vector θ_*, at which setting the model in some way optimally identifies the plant. In the present instance, where it is possible for the model to mimic the plant exactly, the only logical choice for θ_* is the plant parameter vector θ_0, and then $\phi_*(k)$ necessarily takes the form stated.

5.1.4 Global Stability of the Output Error Algorithm

A crucial feature of the arrangement of Figure 5.4, whose stability we now consider, is that it comprises two passive systems connected in a feedback loop, with one of these systems defined by a strictly positive real operator. Let us verify the passivity of the lower block, defined by

$$\theta(k+1) = \theta(k) + \phi(k)m(k) \tag{1.39a}$$

$$p(k) = \phi^T(k)\theta(k+1) \tag{1.39a}$$

We have to show that for all $m(\cdot)$ and K there holds

$$\sum_0^K p(k)m(k) \geq \text{function of } \theta(0) \tag{1.40}$$

Now

$$\begin{aligned} p(k)m(k) &= \theta^T(k+1)\phi(k)m(k) \\ &= \theta^T(k+1)[\theta(k+1)-\theta(k)] \\ &= 1/2\|\theta(k+1)-\theta(k)\|^2 \\ &\quad + 1/2\|\theta(k+1)\|^2 - 1/2\|\theta(k)\|^2 \end{aligned}$$

Hence

$$\sum_0^K p(k)m(k) \geq -1/2\|\theta(0)\|^2$$

and the passivity property is established.

From this fact and the strict positive real nature of $H(q^{-1})$, it follows by standard theory (see Chapter 2) that $v(k) \in l_2$, and so $v(\cdot) \in l_\infty$, $v(k) \to 0$ as $k \to \infty$. Recalling that $v(k)$ is just $C(q^{-1})[y(k)-z(k)]$ [see (1.29)] and that $C(\cdot)$ is stable, it follows that $|y(k)-z(k)| \to 0$ and with very little extra work, $y(k)-\hat{y}(k) \to 0$. This means that convergence of the output errors is guaranteed.

Because $v(\cdot)$ is bounded, as is $\phi_*(\cdot)$, it follows that $\phi(\cdot)$ is bounded [see Figure 5.4 and recall that we established that $H_{\phi v}(q^{-1})$

is stable]. By virtue of the results established in Chapter 2 (Section 2.3), we know that in addition to securing convergence of the output error, we will have exponential convergence of $\theta(k)$ to zero if for some positive α_1, integers S and J, and all $j \geq J$,

$$\sum_{k=j}^{j+S} \phi(k)\phi^T(k) > \alpha_1 I \tag{1.41}$$

Since $|y(k)-z(k)| \to 0$, we have $\|\phi(k)-\phi_*(k)\| \to 0$ as $k \to \infty$ and so (1.41) is equivalent to

$$\sum_{k=j}^{j+S} \phi_*(k)\phi_*^T(k) > \alpha_1 I \tag{1.42}$$

for some α_1 integers S and J, and all $j \geq J$. This in turn is guaranteed by condition (1.15), with the same argument as can be used in the equation error case applying here.

5.1.5 Parameter Estimation When Exact Matching Is Impossible

We turn now to consider the situation that the plant can be modelled approximately by

$$A^*(q^{-1})y^*(k) = q^{-d}B^*(q^{-1})u(k) \tag{1.43}$$

with $A^*(\cdot)$ satisfying the earlier listed conditions on $A(\cdot)$. Moreover, the plant is too complex to permit modelling by a linear transfer function $q^{-d}B(q^{-1})A^{-1}(q^{-1})$ with degrees of A and B as earlier, and (1.43) is the tuned model of the plant in that it is some type of good approximation to the plant, for which we would expect $|y(k)-y^*(k)|$ to be small. If we run the same algorithms as before, and define $\theta(k) = \theta_* - \hat{\theta}(k)$, the parameter error vector for the tuned model of the plant, we find, following the pattern of the earlier calculations, that

$$\begin{aligned}\theta(k+1) &= \theta(k) - \mu\phi(k)C(q^{-1})[y(k)-y^*(k)] \\ &= \theta(k) - \mu\phi(k)H(q^{-1})[\phi^T(k)\theta(k+1)]\end{aligned} \tag{1.44}$$

The parallels with (1.20) are as before [set out just below (1.36)] except that now the parallel of e_* is not zero, but rather $e_*(t) \leftrightarrow$

$$e_*(k) \;=:\; -\mu C(q^{-1})[y(k)-y^*(k)] \tag{1.45}$$

(see Figure 5.4 also). Because $e_*(k) \neq 0$, we cannot expect that $\theta(k) \to 0$.

Specifically, for the output error schemes, one can think of (1.44) as a variant on (1.36), obtained by adding an input term. Then the theory of total stability (Chapter 1) says that if (1.36) is uniformly asymptotically stable, and $e_*(k)$ is small enough, then (1.44) will have a bounded solution. [Strictly, the theory of total stability requires us to work with state variable equations; however, (1.36) is equivalent to a set of state-variable equations, so the idea of total stability can be carried over.] Of course, if $\phi(k)$ is persistently exciting, and $C(q^{-1})[A^*(q^{-1})]^{-1}$ strictly positive real, the premise is fulfilled. Since we no longer have $e_*(k) \equiv 0$, as reference to Figure 5.4 shows, we will not have $\theta(k) \to 0$ or $v(k) \to 0$ and hence we can no longer establish that $\phi(k) \to \phi_*(k)$; however, if we know that as $k \to \infty$, both $e_*(k)$ and $\|\theta(k)\|$ become small, we can conclude that $\|\phi(k)-\phi_*(k)\|$ becomes small, and so a persistence of excitation condition on ϕ_* becomes a persistency of excitation condition on ϕ, and all is well.

The formalization of these arguments is somewhat intricate. See, for example, Anderson and Johnson (1982b) for details. Among the more detailed conclusions is that the allowed bound on $e_*(k)$ depends on the size of the initial parameter error $\theta(0)$; if this is large, the allowed bound on $e_*(\cdot)$ becomes smaller.

Of course, if $C(q^{-1})[A^*(q^{-1})]^{-1}$ is no longer strictly positive real, we have a new problem. Rather than considering the effects here, we shall rely on our treatment of the control problem later in this chapter to capture those ideas which will also apply to the identification problem.

5.2 ADAPTIVE CONTROL ALGORITHM

5.2.1 Model Reference Control for a Known Plant with Possibility of Exact Matching

To obtain insight into the general problem considered in this chapter, we shall consider first problems involving a known plant. Thus suppose the plant is described by

$$A(q^{-1})y(k) = q^{-d}B(q^{-1})u(k) \tag{2.1}$$

where $u(\cdot), y(\cdot)$ are the input and output, $A(0) = 1, B(0) \neq 0$, q^{-1} is the unit delay operator. Later, we shall introduce a modification in the form of a disturbance signal, which can be used to account for nonlinearity, etc. The plant may not be stable in open loop.

The plant is to be controlled so that its response matches, at least as far as possible, the output $z(\cdot)$ of a known stable reference model:

$$C(q^{-1})z(k) = q^{-\bar{d}}D(q^{-1})r(k) \tag{2.2}$$

where $r(\cdot)$ is the reference model input, assumed bounded, and $C(0) = 1$, $D(0) \neq 0$. This means in effect that $u(\cdot)$ can be generated from any combination of $y(\cdot), z(\cdot)$ and $r(\cdot)$, so long as $z(k)-y(k) \to 0$ for all initial conditions and reference signals $r(\cdot)$. Stability of the closed loop system is therefore required.

Another requirement is that $\bar{d} \geq d$; for if $d > \bar{d}$, the plant will have greater delay than the model and it will be impossible for the plant to respond to a step input as fast as the model, without introducing non causal control action.

Another approach to the problem of securing tracking is as follows: let $M(q^{-1}), N(q^{-1})$ be polynomial in q^{-1} and satisfying

$$C(q^{-1}) = M(q^{-1})A(q^{-1}) + q^{-d}N(q^{-1}) \tag{2.3}$$

Because $A(0) \neq 0, A(q^{-1})$ and q^{-d} are coprime, such $M(\cdot), N(\cdot)$ can always be found. We can force uniqueness by demanding additionally that

$$\deg N < \deg A \tag{2.4}$$

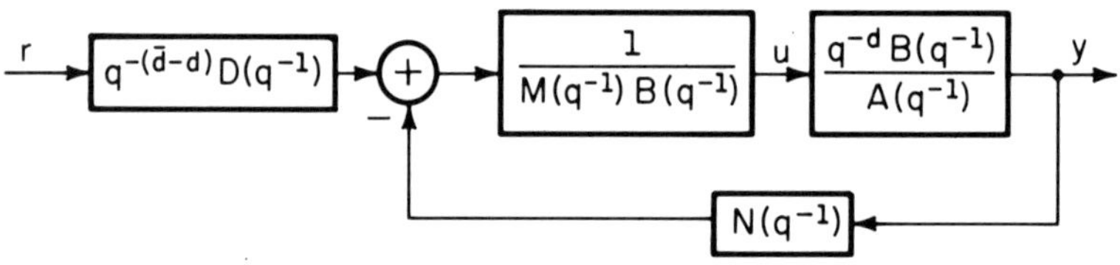

Fig. 5.5 Model matching control.

Now consider the arrangement of Figure 5.5. Under the assumption that $B(q^{-1})$ is stable, it is easily verified that the whole arrangement is stable, with transfer function from r to y of $q^{-\bar{d}}D(q^{-1})C^{-1}(q^{-1})$. Thus asymptotically, y tracks z. The assumptions gathered together are

(A.1) The plant is defined by (2.1), with $A(0) = 1, B(0) \neq 0$

(A.2) The reference model is defined by (2.2) with $C(0) = 1$

(A.3) $C(\cdot)$ is stable

(A.4) $\bar{d} \geq d$

(A.5) $B(\cdot)$ is stable

We have earlier justified (A.4); assumption (A.3) is clearly reasonable a priori, although admittedly one could imagine consideration of problems where (A.3) does not hold; concerning

(A.5), one can ask whether it arises as an artifact of the solution method, or is fundamental. As it turns out, (A5) is fundamental -- in the exact matching case -- as it excludes possible unstable pole zero cancellations. A proof would take us too far afield. When exact matching is impossible, we relax this condition.

With

$$W(q^{-1}) = M(q^{-1})B(q^{-1}) \tag{2.5}$$

Figure 5.5 is evidently defining the control law

$$W(q^{-1})u(k) = q^{-(\bar{d}-d)}D(q^{-1})r(k)-N(q^{-1})y(k) \tag{2.6}$$

Setting

$$W(q^{-1}) = \sum_{i=0}^{\omega} w_i q^{-i} \; ; \; N(q^{-1}) = \sum_{j=0}^{\nu} n_j q^{-j} \; ; \; D(q^{-1}) = \sum_{t=0}^{\delta} d_t q^{-t} \tag{2.7}$$

this can be written as

$$u(k) = -w_0^{-1}[\sum_{i=1}^{\omega} w_i u(k-i)+\sum_{j=0}^{\nu} n_j y(k-j) \tag{2.8}$$

$$-\sum_{t=0}^{\delta} d_t r(k-t-\bar{d}+d)]$$

This is a form to which we shall return.

It will be readily recognized that this control law is, in effect, implementing a *deadbeat* observer. Such observers are notoriously sensitive to output measurement noise, and accordingly, we shall now consider a modification to the above solution procedure which introduces observer dynamics. Later, we shall carry over the introduction of observer dynamics to the adaptive problem, where we should presume that the filtering effect will assist in obtaining robustness.

Consider Figure 5.6 in the following discussion. Let $P(q^{-1}) = \sum_{i=0}^{\pi} p_i q^{-i}$, $p_0 = 1$ denote the observer characteristic polynomial, and define M,N by

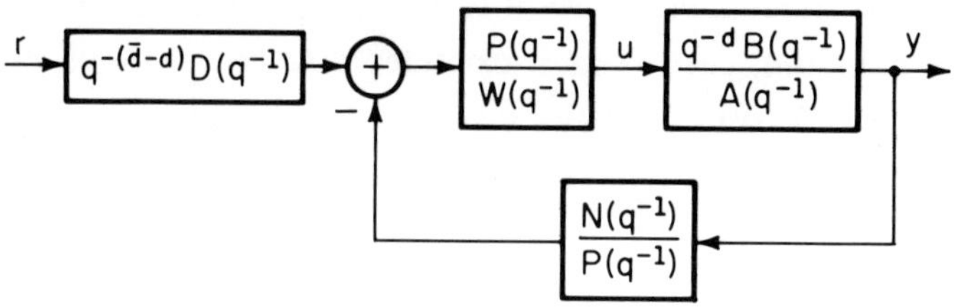

Fig. 5.6 (a) Model matching control with filtering.

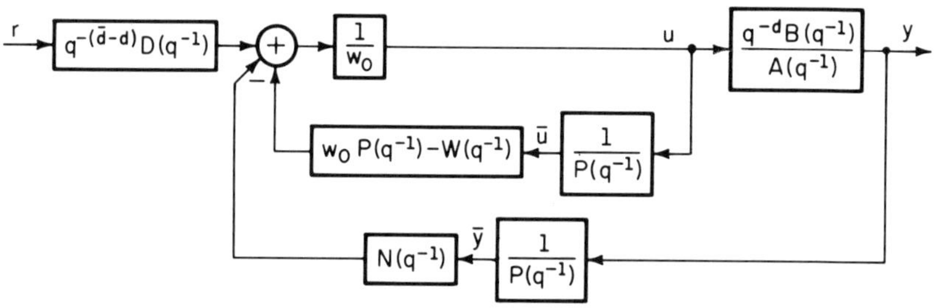

Fig. 5.6 (b) Model matching control with filtering.

$$P(q^{-1})C(q^{-1}) = M(q^{-1})A(q^{-1}) + q^{-d}N(q^{-1}) \tag{2.9}$$

and (2.4). Retain the definition (2.5) of $W(q^{-1})$, but with the new $M(q^{-1})$ naturally.

It is readily verified that the arrangement of Figure 5.6 again provides a transfer function from r to y of $q^{-\bar{d}}D\,(q^{-1})\,C^{-1}(q^{-1})$ with internal stability provided that not just $B(q^{-1})$ is stable, but also $P(q^{-1})$. So to (A.1) through (A.5), we add

(A.6) $P(q^{-1})$ with $P(0) = 1$ is stable

Of course, it should be expected that the observer dynamics have to be stable.

In the material that follows, the question arises of the degrees which M,N may have. Let

$$\begin{aligned} \deg A &= \alpha\,, \quad \deg B = \beta \\ \deg C &= \gamma\,, \quad \deg M = \mu = \omega - \beta \end{aligned} \tag{2.10}$$

Then (2.4) forces

$$\nu \leq \alpha - 1 \tag{2.11}$$

and generically equality will occur. Because the maximum degree on both sides of (2.9) is the same, we will have

$$\pi + \gamma \leq \max(\omega-\beta+\alpha,\, d+\nu) \tag{2.12}$$

and strict inequality can only occur if $\omega-\beta+\alpha = d+\nu$. Finally, the number of adjustable coefficients in $M(q^{-1})$ and $N(q^{-1})$ is $\omega-\beta+\nu+1$ and generically, this must at least equal the number of coefficients of positive powers of q^{-1} on the left side of (2.9), i.e.,

$$\pi + \gamma \leq \omega - \beta + \nu + 1 \tag{2.13}$$

In the generic case, when (2.11) applies with equality, (2.12) and (2.13) can be simplified to

$$\omega - \beta + \alpha = \max(d+\alpha-1,\, \pi + \gamma)$$

Hence generically, we can use

$$\nu = \alpha-1\,, \quad \omega = \max(d+\beta-1,\, \pi+\gamma+\beta-\alpha) \tag{2.14}$$

Finally, we record the control law equation. As reference to Figure 5.6b shows, (2.8) is modified to read

$$\begin{aligned} u(k) = {} & -w_0^{-1}\Big[\sum_{i=1}^{\omega} w_i\bar{u}(k-i)+\sum_{j=0}^{\nu} n_j\bar{y}(k-j) \\ & - \sum_{t=0}^{\delta} d_t r(k-t-\bar{d}+d)\Big] + \sum_{s=1}^{\pi} p_s\bar{u}(k-s) \end{aligned} \tag{2.15}$$

where

$$P(q^{-1})\bar{u}(k) = u(k) \tag{2.16a}$$

or

$$\bar{u}(k) = u(k)-\sum_{s=1}^{\pi} p_s\bar{u}(k-s) \tag{2.16b}$$

and

$$P(q^{-1})\bar{y}(k) = y(k) \tag{2.17}$$

5.2.2 Adaptive Model Reference Control with Possibility of Exact Matching

We move now to a situation in which the coefficients of the polynomial A, B in (2.1) *are unknown*. We assume still that the goal is for the plant output to track the output of the reference model (2.2), and that the assumptions (A.1)-(A.6) in the previous subsection all hold. Thus we will work with a non-deadbeat observer.

The reference model *and* the observer characteristic polynomial are known. The coefficients of the polynomials W,N are not known and must be estimated. Of course, we must assume known the degrees ω and ν of these polynomials, and the values are set using (2.14).

Excluding the part of the overall adaptive system which adjusts the compensator coefficients, the plant and controller are depicted in Figure 5.7. Of course, our task is to find a mechanism for adjusting the coefficents so that stability is retained, and in the end $y(k)-z(k) \to 0$. To this end, we need a regression vector depending on the measurements, and an error (which can be measured) which equates to an inner product of the regression vector and a parameter error vector. We also need a scheme for adjusting our estimate of the parameter using this error.

The parameter vector is selected in an obvious manner. Take

$$\theta_0 = [w_0 \ w_1 \ \cdots \ w_\omega \ n_0 \ n_1 \ \cdots \ n_\nu]^T \tag{2.18}$$

this being the "correct" but initially unknown value for the compensator gain vector. Define also the parameter estimate, available at time k:

$$\hat{\theta}(k) = [\hat{w}_0(k) \cdots \hat{w}_\omega(k) \ \hat{n}_0(k) \cdots \hat{n}_\nu(k)] \tag{2.19}$$

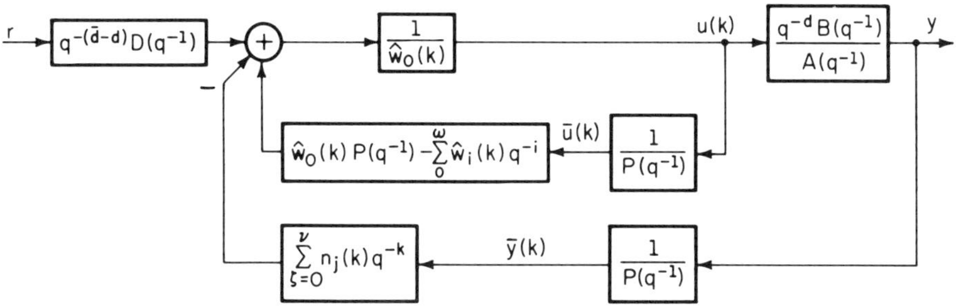

Fig. 5.7 Adaptive model matching control.

For the regression vector, it proves helpful to work with

$$\phi(k) = [\bar{u}(k)\ \bar{u}(k-1) \cdots \bar{u}(k-\omega)\ \bar{y}(k) \cdots \bar{y}(k-\nu)]^T \tag{2.20}$$

How now can we obtain an error? We have as an identity from (2.9) that $PC = MA + q^{-d}N$ and so

$$P(q^{-1})C(q^{-1})y(k) = M(q^{-1})A(q^{-1})y(k) + q^{-d}N(q^{-1})y(k)$$

or

$$P(q^{-1})C(q^{-1})y(k) = q^{-d}M(q^{-1})B(q^{-1})u(k) + q^{-d}N(q^{-1})y(k)$$

or

$$\begin{aligned} C(q^{-1})y(k) &= q^{-d}[W(q^{-1})\bar{u}(k) + N(q^{-1})\bar{y}(k)] \\ &= \phi^T(k-d)\theta_0 \end{aligned} \tag{2.21}$$

Thus although θ_0 is not initially known, we do know at any instant of time k the value of $\phi^T(k-d)\theta_0$, from measurement of $y(k)$ and processing of the latter through $C(q^{-1})$. Since we also know at time k the value of $\phi^T(k-d)\hat{\theta}(k)$, or for that matter $\phi^T(k-d)\hat{\theta}(k-d)$, we can if we desire, form an error from the measurements as

$$e(k) = C(q^{-1})y(k) - \phi^T(k-d)\hat{\theta}(k) \tag{2.22}$$

and we note that in view of (2.21), we have in effect formed

$$e(k) = \phi^T(k-d)[\theta_0 - \hat{\theta}(k)] \tag{2.23}$$

Let us simply set

$$\theta(k) = \theta_0 - \hat{\theta}(k) \tag{2.24}$$

as the parameter error vector.

The question now arises as to how we might update the estimate of $\hat{\theta}(k)$, using the information that $e(k)$ takes a certain value. A standard gradient algorithm could be used, as

$$\hat{\theta}(k+1) = \hat{\theta}(k)+\epsilon(k)\phi(k-d)[C(q^{-1})y(k)-\phi^T(k-d)\hat{\theta}(k)] \tag{2.25}$$

where $\epsilon(k)$ is some, possibly variable, parameter for step-size adjustment, with $0 < \epsilon_{\min} \le \epsilon(k) \le \epsilon_{\max} < \infty$ for all k. A second possibility is to use a *normalized* gradient algorithm, which is

$$\hat{\theta}(k+1) = \hat{\theta}(k) + \frac{\epsilon(k)\phi(k-d)}{1+\epsilon(k)\phi^T(k-d)\phi(k-d)} \times [C(q^{-1})y(k)-\phi^T(k-d)\hat{\theta}(k)] \tag{2.26}$$

The distinction appears trivial. There is nevertheless an advantage in using the latter: it allows a stability proof, as will be clarified below. Yet a third possibility is to use some form of least squares algorithm. For example, least squares with exponential forgetting gives us

$$\hat{\theta}(k+1) = \hat{\theta}(k) + \frac{\rho P(k)\phi(k-d)}{\lambda+\rho(k)\phi^T(k-d)P(k)\phi(k-d)} \times [C(q^{-1})y(k)-\phi^T(k-d)\hat{\theta}(k)] \tag{2.27a}$$

and

$$P(k+1) = \lambda^{-1}\left[I - \frac{\rho P(k)\phi(k-d)\phi^T(k-d)}{\lambda+\rho\phi^T(k-d)P(k)\phi(k-d)}\right]P(k) \tag{2.27b}$$

with $\rho \in (0,1), \lambda \in (0,1), P(0) = \nu I$ for some $\nu > 0$. Least squares with covariance resetting is achieved by

$$\hat{\theta}(k+1) = \hat{\theta}(k) + \frac{\rho P(k)\phi(k-d)}{1+\phi^T(k-d)P(k)\phi(k-d)} \times [C(q^{-1})y(k) - \phi^T(k-d)\hat{\theta}(k)] \tag{2.28a}$$

$$P(k+1) = \left[I - \frac{P(k)\phi(k-d)\phi^T(k-d)}{1+\phi^T(k-d)P(k)\phi(k-d)} \right] P(k) \tag{2.28b}$$

for $k \neq mK$, $m = 0,1,2,3,\ldots$, for K fixed, and

$$P(mK) = \eta_m I \tag{2.29}$$

for $0 < \eta_{\min} \leq \eta_m \leq \eta_{\max} < \infty$. Often, $\rho = 1$ in (2.28a). These sorts of variations in the parameter estimate update algorithm parallel those for identification. We shall focus here on the normalized gradient algorithm of (2.26), as it is most easily understood.

We have left open the value to be assigned to $\epsilon(k)$ in (2.26), and two comments are in order. First, observe that the control law which we shall be implementing, obtainable from (2.15) by using adjustable parameters in lieu of fixed ones, and illustrated in Figure 5.4, is

$$u(k) = -\frac{1}{\hat{w}_0(k)}\left[\sum_{i=1}^{\omega} \hat{w}_i(k)\bar{u}(k-i) + \sum_{j=0}^{\nu} \hat{n}_j(k)\bar{y}(k-j) - \sum_{t=0}^{\delta} d_t r(k-t-\bar{d}+d)\right] + \sum_{s=1}^{\pi} p_s \bar{u}(k-s) \tag{2.30}$$

Evidently, we require $\hat{w}_0(k)$ to be nonzero. One way we can ensure that this holds for all k is to *vary the value of* $\epsilon(k-1)$ *if the calculation of* $\hat{\theta}(k)$ *(before the variation) leads to* $\hat{w}_0(k) = 0$. The second comment to make is: if $\epsilon(k)$ is small, then the variation of $\hat{\theta}(k)$ will be slow, and this may have theoretical advantages (e.g., opening the possibility of use of averaging theory) and practical advantages (e.g., smoothing out the effects of noise).

So far, we have established nothing concerning the stability of the adaptive scheme proposed above, and we have not exhibited the overall scheme in a form similar to that discussed in the previous chapter for continuous-time systems. This will be done in Section 5.3. Meanwhile, we set up a framework to allow tackling of the problem of developing an algorithm when exact matching is not possible.

There is an informative reinterpretation of (2.30) which we shall now derive. Recalling from (2.16a) that $P(q^{-1})\bar{u}(k) = u(k)$, (2.30) becomes equivalent to

$$\hat{w}_0(k)\bar{u}(k) = -\sum_{i=1}^{k}\hat{w}_i(k)\bar{u}(k-i)-\sum_{j=0}^{v}\hat{n}_j(k)\bar{y}(k-j)+q^{d-\bar{d}}D(q^{-1})r(k)$$

or

$$\phi^T(k)\hat{\theta}(k) = C(q^{-1})z(k+d) \tag{2.31a}$$

or

$$\phi^T(k-d)\hat{\theta}(k-d) = C(q^{-1})z(k) \tag{2.31b}$$

[This equation can be regarded as implicitly defining $u(k)$.] Since we know that $\phi^T(k)\theta_0 = C(q^{-1})y(k)$, the interpretation is as follows. At time $k-d$, act as if the current estimate of θ_0, viz., $\hat{\theta}(k-d)$, is correct, and choose $u(k-d)$ so as to ensure that $y(k)$ [the value of which is affected by $u(k-d)$] takes the desired value of $z(k)$.

The reinterpretation in turn gives further insight into the error signal driving the adaptive algorithm, viz. $e(k) = C(q^{-1}) \times [y(k)-\phi^T(k-d)\hat{\theta}(k)]$. This may be regarded as

$$e(k) = C(q^{-1})[y(k)-z(k)]+\phi^T(k-d)[\hat{\theta}(k-d)-\hat{\theta}(k)] \tag{2.32}$$

In a very crude sense, one could say that the difference $e(k)$ measures the difference $y(k)-z(k)$. The slower the rate of change of $\hat{\theta}(k)$, the more accurate is the statement. In continuous-time, the distinction between $e(k)$ and $C(q^{-1})[y(k)-z(k)]$ no longer exists, and matters are perhaps more transparent.

5.2.3 Adaptive Model Reference Control with No Exact Matching Possible

We now turn to the situation where we restrict the controller dimension so that exact asymptotic tracking, i.e., $|z(k)-y(k)| \to 0$, is impossible. We suppose that this is because the true dimension of the plant, which is still defined by (2.1) with $\deg(A) = \alpha$ and $\deg(B) = \beta$, is larger than what we care to try to model.

Let us suppose however that there exists a *tuned* stabilizing controller, defined by polynomials $W^*(q^{-1})$ and $N^*(q^{-1})$, such that with this controller, the closed loop is stable. In what sense, we might ask, is the controller tuned? No totally straightforward or comprehensive answer can be advanced, as noted in Chapter 1. A partial answer is obtainable as follows. We wish $y(k)$ to track $z(k)$. Let us therefore define an error measure by

$$J_1 = \lim_{L \to \infty} L^{-1} \sum_{0}^{L-1} [y(k)-z(k)]^2 \tag{2.33}$$

or even

$$J_2 = \lim_{L \to \infty} L^{-1} \sum_{0}^{L-1} \{H(q^{-1})[y(k)-z(k)]\}^2 \tag{2.34}$$

where $H(q^{-1})$ can be chosen so as to weight selectively different frequency components. *If* there is a fixed stabilizing controller for the plant which achieves minimization of J_1 or J_2 we could regard it as the tuned controller. Of course, its existence and defining parameter values depend on the sequence $r(\cdot)$.

The degrees ω, ν of W^*, N^* must be prespecified in advance. The coefficient values in the adaptive problem are not known. The degrees may be those which would allow exact solution of the tracking problem were the plant of some known lesser order than the actual plant, i.e., W^*, N^* may define the correct controller (correct meaning exact matching is achieved) for a plant approximating the actual plant.

The other assumptions made are (A1) through (A4). We no longer *explicitly* require $B(q^{-1})$ to be stable; but note that if, for example, ω and ν are such that exact matching is possible, then the requirement that the closed loop be stable with no hidden modes implies that $B(q^{-1})$ must be stable. In principle, though, once exact matching is not secured, $B(q^{-1})$ does not have to be stable (see Chapter 3 for an example).

The adaptive algorithm and the generation of the control proceed just as in the last subsection. That is:

- the regression vector is still given by (2.20);
- the parameter vector defining the compensator gain is still given by (2.26);
- the control at time k is still given by (2.30).

But note that it is no longer the case that we have $C(q^{-1})y(k) = \phi^T(k-d)\theta_0$ or $C(q^{-1})y(k) = \phi^T(k-d)\theta_*$. So the justification for using (2.26) is indirect: because it works, i.e., can be justified, in the case when exact matching is possible, we can try it in the inexact matching case (and see, or prove, what happens).

When we come to work with an error model, it turns out that one helpful way to think about this scheme is as follows. Referring to Figure 5.8, define the signal from the output of a tuned controller driven by the regression vector associated with adaptive controller use as

$$\begin{aligned} u_*(k) \;=\; & -\frac{1}{w_0^*}\Big[\sum_{i=1}^{\omega} w_i^*(k)\bar{u}(k-i) + \sum_{j=0}^{\nu} n_j^*(k)\bar{y}(k-j) \\ & - \sum_{l=0}^{\delta} d_l r(k-h-\bar{d}+d)\Big] + \sum_{j=1}^{\pi} p_j \bar{u}(k-j) \end{aligned} \tag{2.35}$$

(Of course, to begin with, the tuned controller is not known to us and so $u_*(k)$ is not known.) The total control can be regarded as the sum

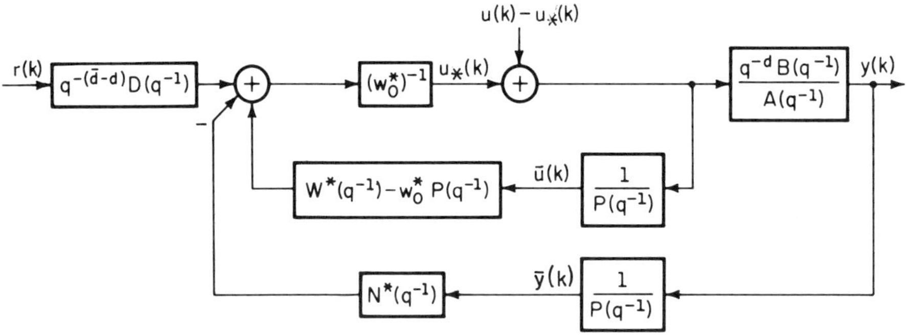

Fig. 5.8 Model matching control; non exact matching situation.

of $u_*(k)$ and a term depending on the parameter error. In this connection, we can calculate the difference $u(k)-u_*(k)$ as follows:

$$
\begin{aligned}
u(k)-u_*(k) \;=\; & \left\{\frac{1}{w_0^*}\Big[\sum_{i=1}^{\infty}[w_i^*-\hat{w}_i(k)]\bar{u}(k-i)+\sum_{l=0}^{\nu}[n_l^*-\hat{n}_l(k)]\bar{y}(k-l)\Big]\right\} \\
& + \left\{\left[\frac{1}{w_0^*}-\frac{1}{\hat{w}_0(k)}\right]\Big[\sum_{i=1}^{\infty}[\hat{w}_i(k)\bar{u}(k-i)+\sum_{l=0}^{\nu}\hat{n}_l(k)\bar{y}(k-l)-\sum_{h=0}^{\delta}d_h r(k-h-\bar{d}+d)\Big]\right\} \\
= \; & \left\{\frac{1}{w_0^*}\phi^T(k)\theta(k) - \frac{w_0^*-\hat{w}_0(k)}{w_0^*}\bar{u}(k)\right\} \\
& + \left\{\left[\frac{1}{w_0^*}-\frac{1}{\hat{w}_0(k)}\right][-\hat{w}_0(k)]\Big[u(k)-\sum_{j=1}^{\pi}p_j\bar{u}(k-j)\Big]\right\} \\
= \; & \frac{1}{w_0^*}\,\phi^T(k)\theta(k)
\end{aligned}
\tag{2.36}
$$

where we are using a new, but obvious, definition of the parameter error as

$$
\theta(k) \;=\; \theta_* - \hat{\theta}(k) \tag{2.37}
$$

In making the above definiton of $u_*(k)$, it is important to stress what it is *not*. If the adaptive scheme is switched off, and a *fixed* controller parametrized by θ_* is used, we will *not* obtain the same $u_*(k)$ sequence for the same input. For as inspection of the figure shows, the value of $u_*(k)$ in part depends on $u(k)-u_*(k)$, which will be absent if the adaptation is turned off. Let us reserve the notation $u^*(k)$ to denote the *tuned control* in this latter case.

5.3 FORMULATION OF AN ERROR MODEL

5.3.1 An Error Model for an Adaptive System Permitting Exact Matching

While our principal interest is in the behavior of an adaptive system with no exact matching possible, we will derive a number of our conclusions for such a system by first considering the exact matching case. Our strategy with this case will be to set up the error equations, and then interpret them. The actual stability analysis is given in the next section.

We adopt all the assumptions listed in Section 5.2.2, and start with the parameter update equation (2.26) repeated as

$$\hat{\theta}(k+1) = \hat{\theta}(k) + \frac{\epsilon(k)\phi(k-d)}{1+\epsilon(k)\phi^T(k-d)\phi(k-d)} \cdot [C(q^{-1})y(k)-\phi^T(k-d)\hat{\theta}(k)] \tag{3.1}$$

Recall the identification of $\phi^T(k-d)\,\theta_0$ from (2.21) repeated as

$$C(q^{-1})y(k) = \phi^T(k-d)\theta_0 \tag{3.2}$$

With parameter error $\theta(k) = \theta_0-\hat{\theta}(k)$, it follows from (3.1) and (3.2) that

$$\theta(k+1) = \left[I - \frac{\epsilon(k)\phi(k-d)\phi^T(k-d)}{1+\epsilon(k)\phi^T(k-d)\phi(k-d)}\right]\theta(k) \tag{3.3}$$

There are two comments that can be immediately made. First, it is apparent from this equation that $\|\theta(k)\|$ is decreasing; for

$$\theta^T(k)\theta(k)-\theta^T(k+1)\theta(k+1) = 2\epsilon(k)\left[\frac{\theta^T(k)\phi(k-d)}{1+\epsilon(k)\phi^T(k-d)\phi(k-d)}\right]^2 \quad (3.4)$$

The remainder of the stability analysis must wait till the next section. Second, (3.3) should *not* be thought of as a linear equation, because $\phi(k-d)$ depends on the settings of the adaptive controller, and thus ultimately on $\theta(\cdot)$.

Of course, (3.3) is not the full story in relation to the stability question. We must consider (3.3), together with the plant equation, the model reference equation and controller law:

$$A(q^{-1})y(k) = q^{-d}B(q^{-1})u(k) \quad (3.5)$$

$$C(q^{-1})z(k) = q^{-\bar{d}}D(q^{-1})r(k) \quad (3.6)$$

and, from (2.30)

$$u(k) = \frac{1}{\hat{w}_0(k)}\Big[\sum_{i=1}^{\omega}\hat{w}_i(k)\bar{u}(k-i) + \sum_{j=0}^{\nu}\hat{n}_j(k)\bar{y}(k-j) - \sum_{t=0}^{\delta}d_t r(k-t-\bar{d}+d)\Big] + \sum_{s=1}^{\pi}p_s\bar{u}(k-s) \quad (3.7)$$

The task is to show that $|y(k)-z(k)| \to 0$ with $u(\cdot)$, $y(\cdot)$ remaining bounded. To establish stability, it actually proves more convenient to work in the next section with the equations

$$C(q^{-1})y(k) = \phi^T(k-d)\theta_0 \quad (3.8)$$

$$C(q^{-1})z(k) = \phi^T(k-d)\hat{\theta}(k-d) \quad (3.9)$$

which were established in (2.21) and (2.31b), respectively.

Let us now interpret the error equations as discrete-time variations of the following continuous-time equations considered in the last chapter of the form

$$e = e_* - H_{ev}v \quad (3.10a)$$

$$\phi = \phi_* - H_{\phi v} v \tag{3.10b}$$

$$v = \phi^T \theta \tag{3.10c}$$

$$\dot{\theta} = \epsilon(\Gamma\phi)e \tag{3.10d}$$

We can identify $\phi(\cdot)$ in the last chapter with $\phi(\cdot)$. Similarly $\theta(\cdot)$ in the last chapter can be identified with $\theta(\cdot)$ in this chapter. In the exact matching case, the tuned error is zero. So the quantity paralleling e_* in the last chapter is, for the moment, zero. The quantity $v(\cdot)$ of the last chapter should be replaced by an inner product of $\phi(\cdot)$ and $\theta(\cdot)$, and H_{ev} becomes the identity operator. We have deliberately omitted mention of time arguments. Actually, a problem of sorts arises since, roughly anyway, $\phi(k-d)$ and $\phi(k)$ can both be regarded as corresponding to $\phi(t)$. To help resolve this problem, let us observe that (3.1), after a little algebra, is equivalent to

$$\begin{aligned} \hat{\theta}(k+1) &= \hat{\theta}(k)+\epsilon(k)\phi(k-d)[C(q^{-1})y(k)-\phi^T(k-d)\hat{\theta}(k+1)] \\ &= \hat{\theta}(k)+\epsilon(k)\phi(k-d)[\phi^T(k-d)\theta_0-\phi^T(k-d)\hat{\theta}(k+1)] \end{aligned}$$

or

$$\theta(k+1) = \theta(k)-\epsilon(k)\phi(k-d)[\phi^T(k-d)\theta(k+1)] \tag{3.11}$$

Notice that (3.11) is like a backward Euler approximation to (3.10d), and suggests the parallels

$$v(t) \leftrightarrow \phi^T(k-d)\theta(k+1)$$

$$\phi(t) \leftrightarrow \phi(k-d)$$

$$e(t) \leftrightarrow \phi^T(k-d)\theta(k+1)$$

$$H_{ev} \leftrightarrow 1$$

$$e_* \leftrightarrow 0$$

$$\epsilon\Gamma \leftrightarrow \epsilon(k)$$

$$\theta(t) \leftrightarrow \theta(k) \text{ or } \theta(k+1) \text{ as appropriate}$$

It is also interesting to note that (3.11) has the form of an *unnormalized* gradient algorithm.

It remains to put down the discrete-time transfer function which is parallel to $H_{\phi v}$, and here the time-discretization problem arises again. Let us define $\bar{v} = \phi^T(k)\theta(k)$. We established in our discussion of the no-matching-possibility case that

$$u(k) - u_*(k) = \frac{1}{w_0^*} \phi^T(k)\theta(k) \tag{3.12}$$

This must remain true in the exact matching case, where $u_*(k)$ refers to the control obtained by the scheme of Figure 5.8. In consequence, we can regard the parallel to $H_{\phi v}$ as the transfer function matrix from $w_0^*[u(k) - u_*(k)]$ to $\phi(k-d)$. This is easily evaluated to be

$$H_{\phi v} = \left[\frac{q^{-d}A}{PBC} \; \frac{q^{-(d+1)}A}{PBC} \; \cdots \; \frac{q^{-(d+w)}A}{PBC} \; \frac{q^{-2d}}{PC} \; \cdots \; \frac{q^{-(2d+v)}}{PC} \right]^T \tag{3.13}$$

5.3.2 An Error Model for an Adaptive System Not Permitting Exact Matching

The core of the adaptive algorithm is still the equation

$$\hat{\theta}(k+1) = \hat{\theta}(k) + \frac{\epsilon(k)\phi(k-d)}{1 + \epsilon(k)\phi^T(k-d)\phi(k-d)}$$
$$\times [C(q^{-1})y(k) - \phi^T(k-d)\hat{\theta}(k)] \tag{3.14}$$

in which it is no longer the case that $C(q^{-1})y(k) = \phi^T(k-d)\theta_*$ for some θ_*. We shall concentrate on the *a priori* error term $e_0(k)$, with

$$-e_0(k) =: C(q^{-1})y(k) - \phi^T(k-d)\hat{\theta}(k) \tag{3.15}$$

[so called because it is based on $\hat{\theta}(k)$ rather than $\hat{\theta}(k+1)$].

Make the definition

$$K(q^{-1}) = \frac{P(q^{-1})B(q^{-1})}{W^*(q^{-1})A(q^{-1}) + q^{-d}N^*(q^{-1})B(q^{-1})} \tag{3.16}$$

The requirement that K is strictly stable (without unstable pole-zero cancellation) is the analogue of assumption (A5) of the exact matching situation. Below, we shall prove the following result.

Lemma 5.1:

For the adaptive algorithm of Section 5.2 (case of exact matching not possible) and with the above definitions of $e_0(\cdot)$ and $K(q^{-1})$, there holds

$$
\begin{aligned}
-e_0(k) &= K(q^{-1})C(q^{-1})[\phi^T(k-d)\theta(k)] \\
&\quad + [K(q^{-1})C(q^{-1})-1][\phi^T(k-d)\hat{\theta}(k)] \qquad (3.17a)\\
&= K(q^{-1})C(q^{-1})[\phi^T(k-d)\theta(k)] \\
&\quad + [K(q^{-1})C(q^{-1})-1]q^{-\bar{d}}D(q^{-1})r(k) \\
&\quad + [K(q^{-1})C(q^{-1})-1]\phi^T(k-d)[\hat{\theta}(k)-\hat{\theta}(k-d)] \qquad (3.17b)
\end{aligned}
$$

Remarks: Before proving the lemma, we wish to lend some significance to $K(q^{-1})$. Refer again to Figure 5.8, and regard the output $y(k)$ as the sum

$$y(k) = y^*(k) + [y(k) - y^*(k)] \qquad (3.18)$$

where the first summand is due to $r(k)$, the second to $u(k)-u_*(k)$. Of course, $y^*(k)$ is the tuned output. In other words, $y^*(k)$ is the output which will result when the controller parameters are set at θ_* and the adaptation is switched off. Then it is not hard to verify that

$$
\begin{aligned}
y^*(k) &= K(q^{-1})q^{-\bar{d}}D(q^{-1})r(k) \qquad (3.19)\\
&= K(q^{-1})C(q^{-1})z(k) \qquad (3.20)
\end{aligned}
$$

while

$$y(k)-y^*(k) = q^{-d}K(q^{-1})w_0^*[u(k)-u_*(k)] \qquad (3.21)$$

Now the tuned parameters are chosen so that $y^*(k)-z(k)$ will be as small as possible. Notice that

$$y^*(k)-z(k) = [K(q^{-1})C(q^{-1})-1]z(k) \tag{3.22}$$

Hence in the frequency range where $z(k)$ has significant energy, we would hope that $|K(e^{j\omega})C(e^{j\omega})-1|$ would be small, or that K will approximate C^{-1}. Equation (3.17a) then suggests that $e(k)$ is approximately $\phi^T(k-d)\theta(k)$ if $KC \simeq 1$ in the relevant frequency range. Equation (3.17b) refines this. The second term of (3.17b) will be small if $|K(e^{j\omega})C(e^{j\omega})-1|$ is small in the frequency range where $r(k)$, or $D(q^{-1})r(k)$ more precisely, has energy, and the third term will be small especially when the adaptive gain $\epsilon(k)$ is kept small, so that the rate of change of $\hat{\theta}(k)$ is small.

In view of the above remarks, it should come as no surprise that in the exact modelling case, where $W = MB$, $MA+q^{-d}NB = C$, one has $KC = 1$.

Proof of Lemma 5.1

The adaptive control system uses the same control law as in the exact matching case. This was given in the previous section as (2.30) (explicit form) or equivalently as (2.31b) (implicit form). Thus (3.20) can be rewritten as

$$y^*(k) = K(q^{-1})\phi^T(k-d)\hat{\theta}(k-d) \tag{3.23}$$

Recalling also the identity (2.36) for $u(k)-u_*(k)$, we obtain with (3.21)

$$y(k)-y^*(k) = K(q^{-1})\phi^T(k-d)\theta(k-d)$$

and so

$$y(k) = K(q^{-1})\phi^T(k-d)\theta_* \tag{3.24}$$

Now from (3.15) we obtain

$$e_0(k) = C(q^{-1})K(q^{-1})\phi^T(k-d)\theta_* - \phi^T(k-d)\hat{\theta}(k)$$

and (3.17a) is immediate.

To obtain (3.17b) from (3.17a), use (2.30) again:

$$\begin{aligned}\phi^T(k-d)\hat{\theta}(k) &= \phi^T(k-d)\hat{\theta}(k-d)+\phi^T(k-d)[\hat{\theta}(k)-\hat{\theta}(k-d)]\\ &= C(q^{-1})z(k)+\phi^T(k-d)[\hat{\theta}(k)-\hat{\theta}(k-d)]\\ &= q^{-\bar{d}}D(q^{-1})r(k)+\phi^T(k-d)[\hat{\theta}(k)-\hat{\theta}(k-d)]\end{aligned}$$

□

The core of the error system now comes from combining (3.14) and (3.17b), and taking $\theta(k) = \theta_*-\hat{\theta}(k)$. It is

$$\begin{aligned}\theta(k+1) = \theta(k)- &\frac{\epsilon(k)\phi(k-d)}{1+\epsilon(k)\phi^T(k-d)\phi(k-d)}\,K(q^{-1})C(q^{-1})\\ &\times\ [\phi^T(k-d)\theta(k)]-\frac{\epsilon(k)\phi(k-d)}{1+\epsilon(k)\phi^T(k-d)\phi(k-d)}\\ &\times\ [K(q^{-1})C(q^{-1})-1]\{q^{-\bar{d}}D(q^{-1})r(k)\\ &-\phi^T(k-d)[\theta(k)-\theta(k-d)]\}\end{aligned} \tag{3.25}$$

This should be compared with the corresponding equation (3.3) for the problem where exact matching is possible. The key differences are:

(a) There are different definitions of parameter error, viz., actual error $\theta_0-\hat{\theta}(k)$, and difference from the "tuned" parameter $\theta_*-\hat{\theta}(k)$.

(b) The homogeneous part of (3.3) has an identity operator replaced by $K(q^{-1})C(q^{-1})$ which in some way should be close to 1.

(c) A forcing term appears in (3.25) where there was none in (3.3). The smaller $K(q^{-1})C(q^{-1})-1$ (via a suitable measure), the smaller is this term. Part of the term can also be made small relative to the part depending on $r(k)$ by reducing the rate of variation of $\hat{\theta}(k)$, i.e., by reducing $\epsilon(k)$.

Now let us consider again how the error system can be regarded as paralleling that of the last chapter. From (3.14), we can derive the following *a posteriori* equation by simple algebra:

$$\hat{\theta}(k+1) = \hat{\theta}(k)+\epsilon(k)\phi(k-d)[C(q^{-1})y(k)-\phi^T(k-d)\hat{\theta}(k+1)] \quad (3.26)$$

which implies

$$\theta(k+1) = \theta(k)-\epsilon(k)\phi(k-d)[C(q^{-1})y(k)-\phi^T(k-d)\hat{\theta}(k+1)] \quad (3.27)$$

This suggests that we identify as the parallel of $e(t)$ the a posteriori error $e(k)$, with

$$\begin{aligned} -e(k) &\triangleq C(q^{-1})y(k)-\phi^T(k-d)\hat{\theta}(k+1) \\ &= K(q^{-1})C(q^{-1})\phi^T(k-d)\theta_*-\phi^T(k-d)\hat{\theta}(k+1) \quad (3.28) \end{aligned}$$

Now the tuned error is the value taken by $e(k)$ when we set $\hat{\theta}(k) = \theta_*$ and turn off the adaptation, i.e., we set $\theta(k) \equiv 0$. This causes a change in the regression vector. Call this changed vector $\phi_*(k)$. Then (3.28) specialized to this situation yields

$$-e_*(k) = (KC-1)\phi_*(k-d)\theta_*$$

Now in the tuned error situation, the equation (2.31a) specializes to $\phi_*^T(k-d)\theta_* = C(q^{-1})z(k)$. But since (2.31a) holds in the untuned adaptive case, it follows that

$$-e_*(k) = (KC-1)\phi^T(k-d)\hat{\theta}(k-d) \quad (3.29)$$

where the quantity on the left is what arises in a tuned case, and the product $\phi^T(k-d)\hat{\theta}(k-d)$ on the right is what arises in the adaptive case. From (3.28) and (3.29), we obtain

$$\begin{aligned} e(k)-e_*(k) &= -\phi^T(k-d)\theta(k+1)-(KC-1)\phi^T(k-d)\theta(k-d) \\ &= -KC\phi^T(k-d)\theta(k+1)-(KC-1)\phi^T(k-d) \\ &\quad \cdot [\theta(k-d)-\theta(k+1)] \quad (3.30) \end{aligned}$$

In terms of parallels to the continuous time equation

$$e = e_* - H_{ev}v \quad (3.31a)$$

$$\phi = \phi_* - H_{\phi v} v \tag{3.31b}$$

$$v = \phi^T \theta \tag{3.31c}$$

$$\dot{\theta} = \epsilon(\Gamma\phi)e \tag{3.31d}$$

we can identify $e(t)$ with $e(k)$, $e_*(t)$ with $e_*(k)$, H_{ev} with KC, $v(t)$ with $v(k) = \phi^T(k-d)\theta(k+1)$. But note that (3.30) is *not* an exact parallel of (3.31a), although we would have in fact, were $\theta(k-d) = \theta(k+1)$,

$$e(k) - e_*(k) = -K(q^{-1})C(q^{-1})v(k) \tag{3.32}$$

which is an exact parallel. This difficulty does not arise in continuous time, because both $\theta(k+1)$ and $\theta(k-d)$ are analogous to $\theta(t)$. As before, the transfer function $H_{\phi v}$ can be computed; the calculation is no different, and one must work into $\bar{v}(k) = \phi^T(k)\theta(k)$ as the input to the transfer function. The parallels of (3.31a), (3.31c), and (3.31d) are shown in Figure 5.9.

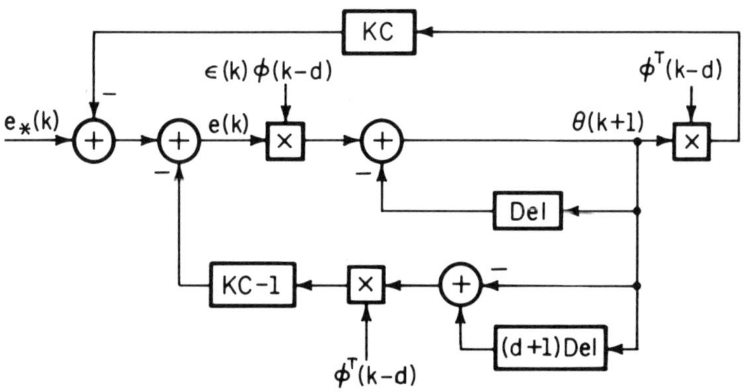

Fig. 5.9 Error model.

Notice that the term $(KC-1)\phi^T(k-d)[\theta(k-d)-\theta(k+1)]$ will be small in one or both of two situations: $|K(e^{j\omega})C(e^{j\omega})-1|$ is small in an appropriate frequency range, and $\hat{\theta}(k)$ is slowly varying [i.e.,

$\epsilon(k)$ is small].

Notice also the following rewriting of $e(k)$ which follows from (3.29), (3.30), and $\phi^T(k-d)\hat{\theta}(k-d) = C(q^{-1})z(k) = q^{-\bar{d}}D(q^{-1})r(k)$:

$$\begin{aligned}-e(k) &= K(q^{-1})C(q^{-1})[\phi^T(k-d)\theta(k+1)] \\ &\quad + [K(q^{-1})C(q^{-1})-1]q^{-\bar{d}}D(q^{-1})r(k) \\ &\quad + [K(q^{-1})C(q^{-1})-1]\phi^T(k-d)[\theta(k-d)-\theta(k+1)] \qquad (3.33)\end{aligned}$$

This can be compared with the *a priori* error formula of (3.17b).

5.4 STABILITY ANALYSIS FOR ADAPTIVE CONTROL

5.4.1 Stability in the Case of Exact Matching

Our focus in this subsection is on the equation

$$\theta(k+1) = \left[I - \frac{\epsilon(k)\phi(k-d)\phi^T(k-d)}{1+\epsilon(k)\phi^T(k-d)\phi(k-d)}\right]\theta(k) \qquad (4.1)$$

[earlier stated as (3.3)], which governs the evolution of the parameter error, together with the boundedness of the signals $u(\cdot)$, $y(\cdot)$ and the convergence to zero of the tracking error $y(k)-z(k)$. We shall make heavy use of

$$C(q^{-1})y(k) = \phi^T(k-d)\theta_0 \qquad (4.2)$$

$$C(q^{-1})z(k) = \phi^T(k-d)\hat{\theta}(k-d) \qquad (4.3)$$

The stability results are given in Properties 1-3 below. The result of Property 1, it will be noted, makes no assumptions on the input sequence $r(k)$ (apart from its boundedness, which was part of the earlier assumptions). On the other hand, Properties 2 and 3 do assume restrictions on the input sequence. It will be the generalization of these last two properties rather than the first with which we will be concerned in considering adaptive systems without the exact matching possibility.

Property 1: For the adaptive scheme described in detail earlier, the sequence $\phi(k)$ is bounded, and $y(k)-z(k) \to 0$.

Proof

We found earlier [see (3.4)] that

$$\theta^T(k)\theta(k)-\theta^T(k+1)\theta(k+1) = 2\epsilon(k)\left[\frac{\theta^T(k)\phi(k-d)}{1+\epsilon(k)\phi^T(k-d)\phi(k-d)}\right]^2$$

from which it is clear that $\|\theta(k)\|$ is decreasing and the sequence

$$\mu(k) = \frac{\theta^T(k)\phi(k-d)}{[1+\epsilon(k)\phi^T(k-d)\phi(k-d)]^{1/2}}$$

approaches zero as $k \to \infty$. From (4.1), it follows that $\|\theta(k+1)-\theta(k)\| \to 0$ as $k \to \infty$, and so

$$v(k) \triangleq \frac{\theta^T(k-d)\phi(k-d)}{[1+\epsilon(k)\phi^T(k-d)\phi(k-d)]^{1/2}} \to 0 \tag{4.4}$$

Now an easy contradiction argument establishes that if, in addition to (4.4),

$$\|\phi(k)\| \leq K_1 \max_{j \leq k} \|\theta^T(j)\phi(j)\| + K_2 \tag{4.5}$$

for some fixed K_1, K_2 and all k, then $|\theta^T(k)\phi(k)| \to 0$. We now verify that (4.5) holds. From (4.2) and (4.3), $\theta^T(k-d)\phi(k-d) = C(q^{-1})[y(k)-z(k)]$ and since $z(\cdot)$ is bounded [$r(\cdot)$ is assumed bounded], (4.5) holds if

$$\|\phi(k)\| \leq K_1' \max_{j \leq k+d} |y(j)| + K_2' \tag{4.6}$$

for all k and some fixed K_1', K_2'.

Because the plant is minimum phase, i.e., $B(q^{-1})$ is stable, there holds

$$|u(k)| \leq K_3' \max_{j \leq k+d} |y(j)| + K_4' \tag{4.7}$$

for all k and some fixed K_3', K_4'. Because of the way $\phi(\cdot)$ is constructed from $y(\cdot)$ and $u(\cdot)$ and because of (4.7), (4.6) follows immediately. [This argument appeared originally in Goodwin, Ramadge, and Caines (1980).] Thus $\theta^T(k)\phi(k) \to 0$. From (4.5) it is immediate that $\phi(k)$ is bounded, while stability of $C(q^{-1})$ and the fact that $\theta^T(k-d)\phi(k-d) = C(q^{-1})[y(k)-z(k)]$ ensure that $|y(k)-z(k)| \to 0$.

Note that Property 1 makes no assertion about the convergence of $\theta(k)$. Also, Property 1 makes no assertions about robustness. Indeed, without further restrictions on $r(\cdot)$, Property 1 is a nonrobust result. For illustration of nonrobust behavior, see, e.g., Anderson (1985).

Of course, robust behavior is obtained by requiring that $\phi(k)$ meets a persistence of excitation type of condition:

Property 2: If for all sufficiently large j and some integer S and real $\sigma_1 > 0, \sigma_2 > 0$ there holds

$$\sigma_1 I < \sum_{j}^{j+S} \phi(k)\phi^T(k) < \sigma_2 I \tag{4.8}$$

then (4.1) exhibits exponential convergence of $\theta(k)$ to zero.

This idea has been well covered earlier in this book. The question now arises: what condition on the external signal $r(\cdot)$ will ensure that $\phi(\cdot)$ possesses the property (4.8)? We assert

Property 3: Define

$$\rho = \max(\pi+\beta+\gamma, d+\alpha+\beta-1) \tag{4.9}$$

If for all sufficiently large j and some integer S and real $\alpha_1 > 0$, $\alpha_2 > 0$ there holds

$$\alpha_1 I < \sum_{k=j}^{j+S-\pi-\beta-\gamma} [r(k)\, r(k+1) \cdots r(k+\rho+\delta)]^T$$

$$\times [r(k)\, r(k+1) \cdots r(k+\rho+\delta)] < \alpha_2 I \tag{4.10}$$

then (4.8) holds for sufficiently large j and (4.1) is exponentially convergent.

This result will be established via several lemmas.

Lemma 5.2:

Consider two bounded sequences $s(k)$, $t(k)$ related by

$$t(k) = E(q^{-1})s(k) \tag{4.11}$$

where $E(q^{-1})$ is a polynomial of degree p in q^{-1} with each coefficient bounded by a constant $\bar{E}$. If for some positive integers S and q, some positive α, and all j, there holds

$$\sum_{k=j}^{j+S} [t(k)\ t(k+1) \cdots t(k+q)]^T [t(k)\ t(k+1) \cdots t(k+q)] > \alpha_1 I \tag{4.12}$$

then

$$\sum_{k=j}^{j+S+p} [s(k)\ s(k+1) \cdots s(k+q)]^T [s(k)\ s(k+1) \cdots s(k+q)] > \alpha_2 I \tag{4.13}$$

for some positive α_2. Second, if for some positive integers S and q, some positive α_3 and all j, there holds

$$\sum_{k=j}^{j+S} [s(k)\ s(k+1) \cdots s(k+p+q)]^T [s(k)\ s(k+1) \cdots s(k+p+q)] > \alpha_3 I \tag{4.14}$$

then

$$\sum_{k=j}^{j+S} [t(k)\ t(k+1) \cdots t(k+q)]^T [t(k)\, t(k+1) \cdots t(k+q)] > \alpha_4 I \tag{4.15}$$

for some positive α_4.

Proof

Suppose that (4.13) fails while (4.12) holds. Then given arbitrary $\delta > 0$, there exists a j and a vector $[a_q \cdots a_o]$ of unit length such that with $a(q^{-1}) = \sum a_i q^{-i}$,

$$|a(q^{-1})s(k)| < \delta \text{ for } k \in [j+q, j+S+p+q]$$

It follows that

$$|E(q^{-1})a(q^{-1})s(k)| < \delta(q+1)\bar{E} \text{ for } k \in [j+p+q, j+S+p+q]$$

or

$$|a(q^{-1})t(k)| < \delta(q+1)\bar{E} \text{ for } k \in [j+p+q, j+S+p+q]$$

By taking δ small enough, a contradiction to (4.12) is obtained. The proof of the reverse implication is similar.

□

If $s(k)$, $t(k)$ are linear combination of sinusoids, e.g.

$$s(k) = \sum_{i=1}^{N} \alpha_i(\exp j\omega_i k) \ , \ \omega_i \neq \omega_j \ , \ \alpha_i \neq 0$$

it is easily established that a condition such as (4.13) holds if and only if $N \geq q+1$. Consequently, the first part of the lemma says that if $t(k)$ contains $(q+1)$ sinusoids, so must $s(k)$, while the second says that if $s(k)$ contains $(p+q+1)$ sinusoids, $t(k)$ must contain $(q+1)$; there is a possibility of the p zeros of $E(q^{-1})$ suppressing p spectral lines at ω_i in $s(k)$.

Lemma 5.3:

With quantities as defined earlier, suppose $A(q^{-1})$, $B(q^{-1})$ are coprime. Then the condition

$$\sum_{k=j}^{j+S-n-\beta} [y(k)\ y(k+1) \cdots y(k+\rho)]^T[y(k)\ y(k+1) \cdots y(k+\rho)] > \alpha_5 I \quad (4.16)$$

for some integer S, positive α_5, and all j implies

$$\sum_{k=j}^{j+S} \phi(k)\phi^T(k) > \alpha_6 I \tag{4.17}$$

for some $\alpha_6 > 0$.

Proof

Suppose that (4.16) holds while (4.17) fails. Then given arbitrary $\epsilon > 0$, there exists a unit length vector $[\lambda_0 \cdots \lambda_\omega \ \kappa_0 \cdots \kappa_\nu]$ such that

$$|\lambda(q^{-1})\bar{u}(k) + \kappa(q^{-1})\bar{y}(k)| < \delta \text{ on } [j,j+S]$$

Here, $\lambda(\cdot),\kappa(\cdot)$ are obviously defined.

This implies

$$|\lambda(q^{-1})P(q^{-1})\bar{u}(k)+\kappa(q^{-1})P(q^{-1})\bar{y}(k)| < \delta(\pi+1)\bar{P} \text{ on } [j+\pi,j+S]$$

where $\bar{P}$ is an overbound on the coefficients of $P(\cdot)$. Equivalently,

$$|\lambda(q^{-1})u(k)+\kappa(q^{-1})y(k)| < \delta(\pi+1)\bar{P} \text{ on } [j+\pi,j+S]$$

so that

$$|\lambda(q^{-1})q^{-d}B(q^{-1})u(k)+\kappa(q^{-1})q^{-d}B(q^{-1})y(k)| < \delta(\pi+1)(\beta+1)\overline{PB}$$

on $[j+\pi+\beta+d,j+S+d]$. Since $A(q^{-1})y(k) = q^{-d}B(q^{-1})u(k)$, this means that with $F(q^{-1}) = \lambda(q^{-1})A(q^{-1}) + \kappa(q^{-1})q^{-d}B(q^{-1})$,

$$|F(q^{-1})y(k)| < \delta(\pi+1)(\beta+1)\overline{PB}$$

on $[j+\pi+\beta+d,j+S+d]$. The degree of F is overbounded by $\max(\omega+\alpha,\nu+\beta+d)$, which is $max(\pi+\beta+\gamma,d+\alpha+\beta-1)$ in the generic case. A contradiction to (4.16) now follows provided we can show that $F(q^{-1}) \neq 0$. Now if $F(q^{-1}) = 0$ then $q^{-d}B(q^{-1})A^{-1}(q^{-1}) = -\lambda(q^{-1})\kappa^{-1}(q^{-1})$, and since $\deg\kappa = \nu < \alpha = \deg A$, we see that B,C are not coprime, which is a contradiction.

□

Notice that Lemma 5.3 relates the persistence of excitation condition on $\phi(\cdot)$ to a similar type of condition, which has been alternatively named sufficiently rich, on $y(\cdot)$. [We remark that earlier, we asserted the existence of an analogous relation between the regressor vector and the input; see the discussion of (1.14) and (1.15). Very similar arguments will establish this relation.] A sufficient condition for (4.16) to hold is that $y(\cdot)$ contains $\rho+1$ different sinusoidal frequencies. The relation from $r(\cdot)$ to $y(\cdot)$ is obtained as follows.

Lemma 5.4:

With quantities as defined earlier, the condition

$$\sum_{k=j}^{j+S-\pi-\beta-\gamma} [r(k)\,r(k+1) \cdots r(k+\rho+\delta)]^T [r(k)\; r(k+1) \cdots r(k+\rho+\delta)] > \alpha_1 I \tag{4.18}$$

for some $\alpha_1 > 0$, some $S > 0$, and all sufficiently large j implies

$$\sum_{k=j}^{j+S-\pi-\beta} [y(k)\,y(k+1) \cdots y(k+\rho)]^T [y(k)\,y(k+1) \cdots y(k+\rho)] > \alpha_5 I \tag{4.19}$$

for some $\alpha_5 > 0$.

Proof

Recognizing that because $y(k)-z(k) \to 0$, (4.19) holds for all sufficiently large j if and only if the same condition holds with $z(k)$ replacing $y(k)$. Then recall that

$$C(q^{-1})z(k) = q^{-\bar{d}}D(q^{-1})r(k)$$

and apply both parts of Lemma 5.2.

The lower bound statements in Property 3 are an immediate consequence of Lemmas 5.3 and 5.4. The upper bound in (4.9) together with Property 1 establishes the upper bound of

(4.8).

□

If $r(\cdot)$ is a linear combination of sinusoids, Property 3 says that exponential convergence will occur if the number of sinusoids is at least $\rho+\delta+1$. In the event that $D(q^{-1})$ has no unit circle zeros, close examination of the earlier arguments will show this number can be reduced to $\rho+1$.

5.4.2 Stability in the Absence of Exact Matching

The central equation we have to study is (3.25), repeated for convenience as

$$\begin{aligned}\theta(k+1) &= \theta(k) - \frac{\epsilon(k)\phi(k-d)}{1+\epsilon(k)\phi^T(k-d)\phi(k-d)} K(q^{-1})C(q^{-1}) \\ &\times [\phi^T(k-d)\theta(k)] - \frac{\epsilon(k)\phi(k-d)}{1+\epsilon(k)\phi^T(k-d)\phi(k-d)} \\ &\times [K(q^{-1})C(q^{-1})-1]\{q^{-\bar{d}}D(q^{-1})r(k) \\ &- \phi^T(k-d)[\theta(k) - \theta(k-d)]\}\end{aligned} \tag{4.20}$$

It is also helpful to contemplate an equivalent *a posteriori* version, obtained from (3.33):

$$\begin{aligned}\theta(k+1) &= \theta(k) - \epsilon(k)\phi(k-d)K(q^{-1})C(q^{-1})[\phi^T(k-d)\theta(k+1)] \\ &- \epsilon(k)\phi(k-d)[K(q^{-1})C(q^{-1})-1]\{q^{-\bar{d}}D(q^{-1})r(k) \\ &- \phi^T(k-d)[\theta(k+1)-\theta(k-d)]\}\end{aligned} \tag{4.21}$$

We shall give two separate treatments of the stability question. The first is shorter, but at one step has to appeal to heuristic reasoning. The second is rigorous, and actually makes more contact with the material of earlier chapters.

The general idea behind the first treatment can be summarized as follows:

- Find conditions for the exponential asymptotic stability of a linearized version of

$$\theta(k+1) = \theta(k) - \frac{\epsilon(k)\phi(k-d)}{1+\epsilon(k)\phi^T(k-d)\phi(k-d)} K(q^{-1})C(q^{-1})$$

$$[\phi^T(k-d)\theta(k)] \tag{4.22}$$

- Require that $|K(e^{j\omega})C(e^{j\omega})-1]|$ is small at those frequencies where $r(\cdot)$ has significant frequency content, and $\epsilon(k)$ is small, so that $\theta(k)-\theta(k-d)$ is small. Then the stability of (4.21) will follow from that of (4.22) by small gain theorems or total stability ideas.

We shall now work through this idea in more detail. Consider first the linear equation

$$\theta(k+1) = \theta(k) - \epsilon(k)\phi_*(k-d)K(q^{-1})C(q^{-1})[\phi_*^T(k-d)\theta(k+1)] \tag{4.23}$$

where $\phi_*(k)$ is the regression vector sequence obtained when the tuned non-adaptive controller is inserted. Note that $\theta(k)$ in (4.23) is *not* supposed to correspond to a parameter error encountered with some interconnection of plant and controller; Equation (4.23) has mathematical significance only. As we know, under a persistence of excitation condition on ϕ_* [and thus on $r(\cdot)$] we can establish exponential stability for (4.23) under either or both of two conditions:

$$K(q^{-1})C(q^{-1}) \text{ is discrete positive real} \tag{4.24}$$

or $\epsilon(k)$ is small and the averaging theory applies. [Of course, to exploit the averaging theory, we must have some perception of the frequency content of $\phi_*(k-d)$. Broadly, we require $r(k)$ to have most of its frequency content where $\text{Re}[K(e^{j\omega})C(e^{j\omega})] > 0$. Since $K(q^{-1})C(q^{-1})$ should be as close to 1 as possible, certainly $\text{Re}[K(e^{j\omega})C(e^{j\omega})] > 0$ in those frequency ranges where $K(e^{j\omega})C(e^{j\omega})$ is close to 1 (cf. Chapter 2, Section 2.4, and Chapter 3, Section 3.5).]

Next, consider the linear equation

$$\begin{aligned}\theta(k+1) &= \theta(k) - \epsilon(k)\phi_*(k-d)K(q^{-1})C(q^{-1})[\phi_*^T(k-d)\theta(k+1)] \\ &- \epsilon(k)\phi_*(k-d)[K(q^{-1})C(q^{-1})-1]\{q^{-\bar{d}}D(q^{-1})r(k) \\ &- \phi_*^T(k-d)[\theta(k+1)-\theta(k-d)]\}\end{aligned} \tag{4.25}$$

The homogeneous part of (4.25) differs from (4.23) by a term on the right side, which is $O[\epsilon^2(k)]$. Accordingly, for $\epsilon(k)$ small enough, the homogeneous version of (4.25) will inherit the exponential stability property of (4.23). The fact that it is not homogeneous means that the solution of (4.25) does not tend to zero as $k \to \infty$, but will be bounded by a quantity depending on

$$M = \|\{[K(q^{-1})C(q^{-1})-1]q^{-\bar{d}}D(q^{-1})r(\cdot)\}\|_\infty \tag{4.26}$$

Assuming that $\phi_*(k-d)$ is bounded, it can be argued that $\phi_*(k-d)-\phi(k-d)$ is $O(\max_{j\le k}\|\theta(j)\|)$, where $\theta(j)$ is the parameter error vector encountered in the adaptive algorithm; it follows that (4.25) is a linearized version of

$$\begin{aligned}\theta(k+1) &= \theta(k) - \epsilon(k)\phi(k-d)K(q^{-1})C(q^{-1})[\phi^T(k-d)\theta(k+1)] \\ &+ \epsilon(k)\phi_*(k-d)[K(q^{-1})C(q^{-1})-1]\{q^{-\bar{d}}D(q^{-1})r(k)\} \\ &- \epsilon(k)\phi(k-d)\phi^T(k-d)[\theta(k+1)-\theta(k-d)]\end{aligned} \tag{4.27}$$

Accordingly, (4.27) will remain bounded-input, bounded-output stable if $\|\theta(0)\|$ is not too great, and M is not large, so that $\max_k\|\theta(k)\|$ is small.

Now (4.27) is almost the same as (4.21); the right hand sides differ by

$$\epsilon(k)[\phi(k-d)-\phi_*(k-d)][K(q^{-1})C(q^{-1})-1]\{q^{-\bar{d}}D(q^{-1})r(k)\}$$

Since, as already noted, $\phi_*(k-d)-\phi(k-d)$ is $O(\max_{j\le k}\|\theta(j)\|)$, our extension (not rigorously justified) of total stability ideas indicates that (4.21) will inherit a local stability property.

The above argument thus suggests that (4.21), or equivalently (4.20), has a bounded solution provided

- $r(\cdot)$ is sufficiently frequency rich to cause the tuned regression vector $\phi_*(\cdot)$ to be persistently exciting;
- $K(q^{-1})C(q^{-1})$ is strictly positive real, or if not, some averaging or other theory ensures the exponential stability of (4.23);
- $\epsilon(k)$ is sufficiently small;
- $|[K(e^{j\omega})C(e^{j\omega})-1]D(e^{j\omega})|$ is small in frequency bands where $r(\cdot)$ has significant power;
- $\theta(0)$ is not large.

We turn now to a more rigorous approach to developing the same conclusions. (As it turns out, one further pointer to circumstances leading to a bounded solution becomes evident.) The key to the second approach is to ensure that we work with a state-variable set of equations, since it is for just this type of equation set that many of the theorems given earlier, especially in Chapter 3, are applicable. The steps we shall follow are:

- Set up a state-variable description of the complete nonlinear adaptive control scheme.
- Linearize in a neighbourhood of a trajectory defined by $\theta(k) \equiv 0$ and $\phi(k) \equiv \phi_*(k)$.
- Apply the combination of singular perturbation and averaging theory ideas of Chapter 3.

Step 1: Obtaining the State-Variable Equations. Let us define

$$Z(k+1) = (\bar{u}(k) \cdots \bar{u}(k-n_1)\, \bar{y}(k) \cdots \bar{y}(k-n_2)) \quad (4.28a)$$

$$n_1 = \beta+d-1 \geq \omega \ , \ n_2 = \alpha-1 \geq \nu \tag{4.28b}$$

These inequalities are consistent with the inexact matching case where we use a controller of too low a dimension to allow exact matching.

$$Y(k+1) = (Z(k+1)^T Z(k)^T \cdots Z(k-d+1)^T)^T \tag{4.28c}$$

$$v(k) = q^{-\bar{d}} D(q^{-1}) r(k) \tag{4.28d}$$

We assert then that there exist matrices $A(\theta(k))$, $B(\theta(k))$, and C such that

$$Y(k+1) = A(\theta(k))Y(k) + B(\theta(k))v(k) \tag{4.29}$$

$$X(k-d) = CY(k) \tag{4.30}$$

This representation in effect follows from the plant equations (2.1), filter equations (2.16) and (2.17), and the control equations (2.30). Indeed, we have:

$$\bar{y}(k) = -\sum_{i=1}^{\alpha} a_i \bar{y}(k-i) + \sum_{i=1}^{\beta} b_i \bar{u}(k-d-i) \tag{4.31}$$

$$\bar{u}(k) = -\frac{1}{\hat{w}_0(k)} \{ \sum_{i=0}^{\omega-1} \hat{w}_{i+1}(k) \bar{u}(k-1-i) + \sum_{i=0}^{\nu} \hat{n}_i(k) \bar{y}(k-i) \} + \frac{1}{\hat{w}_0(k)} v(k) \tag{4.32}$$

Substituting (4.31) into (4.32), we obtain:

$$\bar{u}(k) = -\frac{1}{\hat{w}_0(k)} \{ \sum_{i=1}^{\omega} \hat{w}_i(k) \bar{u}(k-i) + \sum_{i=1}^{\beta} b_i \hat{n}_0(k) \bar{u}(k-d-i) \}$$

$$+ \sum_{i=1}^{\alpha} a_i \hat{n}_0(k) \bar{y}(k-i) - \sum_{i=1}^{\nu} \hat{n}_i(k) \bar{y}(k-i) \} + \frac{1}{\hat{w}_0(k)} v(k) \tag{4.33}$$

From (4.31) and (4.33) and (4.28) the representation (4.29), (4.30) is immediate. However, this is not the complete story, as the update equations for $\theta(k)$ need also to be brought into the picture. Using (F_1,G_1,H_1,D_1) as a state-space realization for $K(q^{-1})C(q^{-1})-1$ and (F_2,G_2,H_2) as a state-space realization for

$$V(q^{-1}) = \operatorname{diag}(q^{-d} \cdots q^{-d}) \quad (\nu+\omega+2 \text{ elements})$$

we can replace (4.20) by a state-space type of equation as follows:

$$\begin{bmatrix} \chi_1(k+1) \\ \chi_2(k+1) \\ \theta(k+1) \end{bmatrix} = \begin{bmatrix} F_1 & G_1 Y^T(k) C^T H_2 & 0 \\ 0 & F_2 & G_2 \\ -\alpha_1(k) H_1 & -\alpha_1(k) D_1 Y^T(k) C^T H_2 & I - \alpha_2(k) Y^T(k) C^T \end{bmatrix}$$

$$\times \begin{bmatrix} \chi_1(k) \\ \chi_2(k) \\ \theta(k) \end{bmatrix} + \begin{bmatrix} G_1 \\ 0 \\ -\alpha_1(k) D_1 \end{bmatrix} \nu(k) \tag{4.34}$$

where

$$\alpha_1(k) = \epsilon(k) \frac{CY(k)}{1+\epsilon(k) Y^T(k) C^T C Y(k)} \tag{4.35}$$

It is (4.29) and (4.34) *together* which constitute the desired state-variable equation set.

Step 2: Linearization of the State-Variable Equation Description. Before describing the linearization procedure in detail, we remind the reader of our discussion in Section 1.3 of Chapter 1 concerning linearization. There we showed how the adaptive system could be linearized about a nominal system state, see, e.g., (1.3.1)-(1.3.5). This procedure can be interpreted as linearizing the trajectories of one system about a fixed trajectory of a second system. Let us make this discussion concrete for the case at hand.

We linearize the combined system (4.29), (4.34) in a neighborhood of a trajectory of the system described by

$$\begin{bmatrix} \bar{Y}(k+1) \\ \bar{\chi}_1(k+1) \\ \bar{\chi}_2(k+1) \\ \bar{\theta}(k+1) \end{bmatrix} = \begin{bmatrix} A(\bar{\theta}(k)) & 0 & 0 & 0 \\ 0 & F_1 & G_1 \bar{Y}^T(k) C^T H_2 & 0 \\ 0 & 0 & F_2 & G_2 \\ 0 & 0 & 0 & I \end{bmatrix} \begin{bmatrix} \bar{Y}(k) \\ \bar{\chi}_1(k) \\ \bar{\chi}_2(k) \\ \bar{\theta}(k) \end{bmatrix} + \begin{bmatrix} B(\bar{\theta}(k)) \\ G_1 \\ 0 \\ 0 \end{bmatrix} \nu(k) \tag{4.36a}$$

Equation (4.36a) is obtained from (4.29) and (4.34) by formally setting $\epsilon(k) \equiv 0$. In fact we pick the trajectory of (4.36a) associated with the initial conditions

$$[\bar{Y}(0)\ \bar{\chi}_1(0)\ \bar{\chi}_2(0)\ \bar{\theta}(0)] = [\bar{Y}(0)\ \bar{\chi}_1(0)\ 0\ 0] \tag{4.36b}$$

Notice that this initial condition ensures that $\bar{\theta}(k) \equiv 0$, and then that $\bar{\chi}_2(k) \equiv 0$. The trajectories being considered are those associated with the use of the tuned controller, and so we elect to replace $\bar{Y}(k)$ by $Y^*(k)$ to emphasize this fact.

The deviations of the trajectories of (4.29)-(4.34) from the above selected trajectory of system (4.36) are denoted as

$$\tilde{Y}(k) = Y(k)-\bar{Y}(k) = Y(k)-Y^*(k)$$

$$\tilde{\chi}_i(k) = \chi_i(k)-\bar{\chi}_i(k) \quad i = 1,2$$

$$\tilde{\theta}(k) = \theta(k)-\bar{\theta}(k) = \theta(k)$$

The linearized equation governing these deviations is

$$\begin{bmatrix} \tilde{Y}(k+1) \\ \tilde{\chi}_1(k+1) \\ \tilde{\chi}_2(k+1) \\ \tilde{\theta}(k+1) \end{bmatrix} = \begin{bmatrix} A(0) & 0 & 0 & \alpha^*(k) \\ 0 & F_1 & G_1 Y^{*T}(k) C^T H_2 & 0 \\ 0 & 0 & F_2 & G_2 \\ -\alpha_0^*(k) & -\alpha_1^*(k) H_1 & -\alpha_1^*(k) D_1 Y^{*T}(k) C^T H_2 & I-\alpha_1^*(k) Y^{*T}(k) C^T \end{bmatrix}$$

$$\begin{bmatrix} \tilde{Y}(k) \\ \tilde{\chi}_1(k) \\ \tilde{\chi}_2(k) \\ \tilde{\theta}(k) \end{bmatrix} + \begin{bmatrix} 0 \\ 0 \\ 0 \\ -\alpha_1^*(k)[H_1 \bar{\chi}_1(k) + D_1 \nu(k)] \end{bmatrix} \tag{4.37}$$

where we have

$$\alpha^*(k) = (\alpha_{ij}^*(k)) \tag{4.38a}$$

$$\alpha_{ij}^*(k) = \sum_m \frac{\partial A_{im}}{\partial \theta_j}\Big|_{\theta=0} Y_m^*(k) + \frac{\partial B_i}{\partial \theta_j}\Big|_{\theta=0} \nu(k) \tag{4.38b}$$

$$\alpha_0^*(k) = [\cdots \epsilon(k) \frac{C_i[H_1 \bar{\chi}_1(k) + D_1 \nu(k)]}{1+\epsilon(k) Y^{*T}(k) C^T C Y^*(k)} \cdots] + O(\epsilon^2(k)) \tag{4.38c}$$

$$\alpha_1^*(k) = \epsilon(k) \frac{CY^*(k)}{1+\epsilon(k)Y^{*T}(k)C^TCY^*(k)} \tag{4.38d}$$

$$C_i = i^{th} \text{ column of } C \tag{4.38e}$$

The benefit from using the state variable description is that it allows unambiguous use of theorems set up for state variable equations. The cost is that some familiar quantities get disguised. For example, the driving term in (4.37) $-\alpha_1^*(k)(H_1\bar{\chi}_1(k)+D_1v(k))$ is what we earlier denoted as

$$\frac{-\epsilon(k)\phi_*(k-d)}{1+\epsilon(k)\phi_*^T(k-d)\phi_*(k-d)}[K(q^{-1})C(q^{-1})-1]v(k) \tag{4.38f}$$

(a fact we shall use below).

For future reference we define

$$\tilde{\chi}(k) = [\tilde{Y}^T(k)\ \tilde{\chi}_1^T(k)\ \tilde{\chi}_2^T(k)]^T$$

and rewrite the homogeneous part of (4.37) as

$$\begin{bmatrix}\tilde{\chi}(k+1)\\ \tilde{\theta}(k+1)\end{bmatrix} = \begin{bmatrix}A_{11}(k) & A_{12}(k)\\ A_{21}(k,\epsilon) & A_{22}(k,\epsilon)\end{bmatrix}\begin{bmatrix}\tilde{\chi}(k)\\ \tilde{\theta}(k)\end{bmatrix} \tag{4.39}$$

where the A_{ij} have the obvious definitions, corresponding to the state partitioning.

The matrix $A_{11}(k)$ defines an exponentially stable state transition matrix. This follows from its triangular structure and because the constant matrices on the main diagonal $(A(0),F_1,F_2)$ are all exponentially stable by assumption. F_2 has only zero eigenvalues, because (F_2,H_2,G_2) is a minimal realization for $V(q^{-1}) = \text{diag}(q^{-d} \cdots q^{-d})$. F_1 is exponentially stable because (F_1,H_1,G_1,D_1) is a minimal realization for $K(q^{-1})C(q^{-1})-1$. Further, a close examination of the procedure used for deriving the state variable equation (4.29) will reveal that the eigenvalues of $A(0)$ are the zeros of the polynomial

$$z^{(n_1+n_2+2)(d+1)}[W^*(z^{-1})A(z^{-1})+z^{-d}N^*(z^{-1})B(z^{-1})]$$

The part in square brackets should at least not be surprising because $\theta(k) \equiv 0$ amounts to using nonadaptive control, and (4.29), (4.30) with $\theta(k) \equiv 0$ describe the plant/controller interaction in the nonadaptive case. The eigenvalues of $A(0)$ define the closed loop poles, modulo delays and possible pole-zero cancellations. Hence $A(0)$ is also exponentially stable.

We remark that Equation (4.39) is in the form suitable to apply Lemma 3.4 of Chapter 3 -- cf. Equations (3.31) in Chapter 3 and (4.39) in this chapter.

Step 3: Introduction of Singular Perturbation and Averaging Ideas. First we establish conditions for the exponential stability of the equation (4.39). This is achieved using the tools prepared in Chapter 3: averaging and singular perturbation. Next, invoking a total stability result of Chapter 1, we conclude that the original nonlinear nonhomogeneous system (4.29)-(4.34) is locally [in the neighborhood of the specified trajectory of system (4.36)] stable. These ideas are now made concrete.

Except for a change of notation Equation (4.39) is like Equation (3.31) in Chapter 3. Lemma 3.4 in Chapter 3 deals with the time scale separation of slow and fast states of Equation (3.31). All hypotheses of Lemma 3.4, Chapter 3, are satisfied by the Equation (4.39); hence we can triangularize equation (4.39) by a Lyapunov transformation of the state [just like the $L-$transformation (3.32) of Chapter 3], to obtain a separation of the states into fast states and the slow parameter $\tilde{\theta}$. This transformation preserves the stability properties and allows us to analyze fast and slow states independently. The fast state equation is immediately seen to be exponentially stable, provided that one restricts ϵ in magnitude. The stability of the slow $\tilde{\theta}$ equation can be analyzed using the techniques of Chapter 3, Section 3.5. which further restrict the magnitude of ϵ in order to guarantee the exponential stability.

The transformation has the form [cf. Equation (3.32) of Chapter 3]

$$\begin{bmatrix} \psi(k) \\ \tilde{\theta}(k) \end{bmatrix} = \begin{bmatrix} I & -L(k,\epsilon) \\ 0 & I \end{bmatrix} \begin{bmatrix} \tilde{\chi}(k) \\ \tilde{\theta}(k) \end{bmatrix} \tag{4.40}$$

where according to Lemma 3.4, $L(k,\epsilon)$ can be approximated up to order ϵ by $L_0(k)$, the steady state solution of

$$L_0(k+1) = A_{11}(k)L_0(k) + A_{12}(k) \tag{4.41}$$

The fast state ψ is governed by the equation (correct up to terms of the order ϵ^2) -- remember that the $L-$transformation decouples fast and slow variables:

$$\psi(k+1) = (A_{11}(k)-L_0(k+1)A_{21}(k,\epsilon))\psi(k) \tag{4.42}$$

Obviously (4.42) is exponentially stable for sufficiently small ϵ, because $A_{11}(k)$ is (see end of previous section).

The homogeneous part of the equation governing the slow state $\tilde{\theta}$ is given by

$$\tilde{\theta}(k+1) = (A_{22}(k,\epsilon) + A_{21}(k,\epsilon)L_0(k))\tilde{\theta}(k) \tag{4.43}$$

In order to investigate its stability properties let us interpret the quantities appearing in (4.43). $A_{22}(k,\epsilon)$ is defined by (4.37) and (4.39):

$$A_{22}(k,\epsilon) = I - \alpha_1^*(k)Y^{*T}(k)C^T$$

which is equivalent to

$$A_{22}(k,\epsilon) = I - \frac{\epsilon(k)\phi_*(k-d)\phi_*^T(k-d)}{1+\epsilon(k)\phi_*^T(k-d)\phi_*(k-d)} \tag{4.44}$$

[cf. Equations (4.38d)].

To understand the second term $A_{21}(k,\epsilon)L_0(k)$ in (4.43) we first interpret the different elements of $L_0(k)$. Therefore we partition $L_0(k)$ according to the partitioning of the $\mathrm{diag}(A_{11}(k)) = (A(0),F_1,F_2)$ [see Equations (4.39), (4.37)] as

$L_0(k) = (L_{01}^T(k)\, L_{02}^T(k)\, L_{03}^T(k))^T$. For the interpretation of $L_{01}(k)$, notice that the sensitivity of the tuned state vector $Y^*(k)$ with respect to the tuned parameter setting, i.e., $\partial Y^*/\partial\theta|_{(\theta=0)}$, satisfies

$$\left(\frac{\partial Y^*}{\partial \theta}\right)\Big|_{\theta=0}(k+1) = A(0)\left(\frac{\partial Y^*}{\partial \theta}\right)\Big|_{\theta=0}(k)+\alpha^*(k)$$

which is obtained by differentiating (4.29) with respect to θ and setting $\theta(k) \equiv 0$. [Recall the definition of $\alpha^*(k)$ in (4.38a) and (4.38b).] Hence we can identify

$$L_{01}(k) = \left(\frac{\partial Y^*}{\partial \theta}\right)\Bigg|_{\theta=0}(k) \tag{4.45a}$$

The third component satisfies the equation

$$L_{03}(k+1) = F_2 L_{03}(k) + G_2$$

Consequently, because the system realized by (F_2,G_2,H_2) copies the input and delays it d sample intervals, $L_{03}(k)$ satisfies

$$H_2 L_{03}(k) \equiv I \tag{4.45b}$$

Using (4.45b) in (4.41) we find that the second component $L_{02}(k)$ satisfies

$$L_{02}(k+1) = F_1 L_{02}(k) + G_2 Y^{*T}(k) C^T$$

Hence using the relation $CY^*(k) = \phi_*^T(k-d)$ and the fact that (F_1,G_1,H_1,D_1) is a realization for $K(q^{-1})C(q^{-1})-1$ we have that

$$H_1 L_{02}(k) = [(K(q^{-1})C(q^{-1})-1)-D_1]\phi_*^T(k-d) \tag{4.45c}$$

Substituting the above expressions (4.45) into $A_{21}(k,\epsilon)L_0(k)$ using the definition of $A_{21}(k,\epsilon)$ from (4.37),(4.39) we have that

$$A_{21}(k,\epsilon)L_0(k) = -\alpha_0^*(k)\left(\frac{\partial Y^*}{\partial \theta}\right)\Bigg|_{\theta=0}$$

$$-\alpha_1^*(k)[(K(q^{-1})C(q^{-1})-1)-D_1]\phi_*^T(k-d)$$

$$-\alpha_1^*(k)D_1\phi_*^T(k-d) \qquad (4.46)$$

Noting that $\alpha_0^*(k)$ in (4.38c) can be rewritten as

$$\alpha_0^*(k) = \epsilon(k)\frac{1}{1+\epsilon(k)\phi_*^T(k-d)\phi_*(k-d)}$$

$$\times\ [K(q^{-1})C(q^{-1})-1]v(k)C$$

where we have used the definition of C_l and $H_1\overline{\chi}_1(k)+D_1v(k)$, and combining this with (4.46) and (4.38d), we finally arrive at the following expression for the homogeneous part of the state equation governing the slow parameter $\tilde{\theta}$:

$$\tilde{\theta}(k+1) = \{I-\epsilon(k)\phi_*(k-d)K(q^{-1})C(q^{-1})\phi_*^T(k-d)-\epsilon(k)\ \frac{\partial \phi_*}{\partial \theta}\Bigg|_{\theta=0}(k-d)$$

$$\times\ [K(q^{-1})C(q^{-1})-1]v(k)\}\tilde{\theta}(k) \qquad (4.47)$$

where we made the simplifying approximation:

$$(1+\epsilon(k)\phi_*^T(k-d)\phi_*(k-d))^{-1} = 1 + O(\epsilon) \qquad (4.48)$$

Notice that, apart from a change in notation, Equation (4.47) is of the form (4.2) in Chapter 3. It suffices to identify the matrix $R(k)$ of Equation (4.2) in Chapter 3 with

$$R(k) = \phi_*(k-d)K(q^{-1})C(q^{-1})\phi_*^T(k-d) + \frac{\partial \phi_*}{\partial \theta}\Bigg|_{\theta=0}(k-d)$$

$$\times\ [K(q^{-1})C(q^{-1})-1]v(k) \qquad (4.49)$$

Theorems 3.1 and 3.5 of Chapter 3, which deal with equations of the general form (4.2), yield conditions under which the stability of (4.47) implies stability of the homogeneous part of (4.37), i.e., (4.39). These Theorems state that there exists a small positive constant ϵ^*, such that for $\epsilon(k) \in (\epsilon_1^*, \epsilon^*)$ $0 < \epsilon_1^* < \epsilon^*$, (4.47) is exponentially stable provided that the matrix $R(k)$ given by (4.49) is "on average" positive; more precisely, provided there exists an ordered sequence of sample times $\{T_k\}$ and a matrix $P = P^T > 0$, such that the sample averages, $\bar{R}(T_k)$, defined as

$$\bar{R}(T_k) = \frac{1}{T_k - T_{k-1}} \sum_{k=T_{k-1}}^{T_k - 1} R(k) \tag{4.50}$$

satisfy the condition

$$P\bar{R}(T_k) + \bar{R}^T(T_k)P > 0 \quad \forall\, k \tag{4.51}$$

This condition can be interpreted as a persistence of excitation condition. In the frequency range of $v(k)$, the transfer function $K(q^{-1})C(q^{-1})$ should be close to 1; then $R(k)$ will be close to $\phi_*(k-d)K(q^{-1})C(q^{-1})\phi_*^T(k-d)$ and condition (4.51) is the usual persistence of excitation condition on the regressor vector. Notice that the realization $(A(0),B(0),C)$ is output reachable and hence a sufficiently rich input $v(k)$ generates a persistently exciting regressor (output) $\phi_*(k-d)$. (See Section 2.5 of Chapter 2.) That $(A(0),B(0),C)$ is output reachable is most easily verified by noticing that the subsystem from $v(k)$ to $Z(k)$ is controllable, and that $\phi(k-d)$ is obtained from $Z(k)$ by d delays in time and deleting some components. The controllability of the subsystem $[v(k)$ to $Z(k)]$ follows from the assumption that the original plant is controllable $[A(q^{-1}),B(q^{-1})$ coprime$]$.

Notice also that it is helpful in order to satisfy (4.51) to have

$$\left.\frac{\partial \phi_*}{\partial \theta}\right|_{\theta=0}(k)$$

small, i.e., the tuned regressor should be insensitive to changes in the tuned parameter at the tuned parameter setting θ_*. This point was *not* brought out explicitly in the earlier heuristic analysis.

Using the Total Stability Theorem 1.4, we can infer from the exponential stability of the linear homogeneous system (4.39) local BIBO stability in a neighborhood of the (bounded) trajectory specified by (4.36) for the nonlinear nonhomogeneous system (4.29), (4.34). This neighborhood has a (guaranteed) radius proportional to the rate of exponential convergence of the linear system; therefore it is sufficient to bound the driving term (4.38f) as

$$\left\| \frac{-\epsilon(\cdot)\phi_*(\cdot-d)}{1+\epsilon(\cdot)\phi_*^T(\cdot-d)\phi_*(\cdot-d)} [K(q^{-1})C(q^{-1})-1]v(\cdot) \right\|_\infty = O(\epsilon^*)$$

Summarizing: The adaptive closed loop system (4.29), (4.34) is locally stable in the neighborhood of the tuned trajectory prescribed by (4.26) provided that the input sequence $v(k) = q^{-\bar{d}}D(q^{-1})r(k)$ is sufficiently exciting and provided that there exists a θ_*, depending on this input sequence, such that:

- $K(q^{-1})C(q^{-1})$ is exponentially stable, and close to 1 in the frequency range of $v(k)$,
- $\phi_*(k-d)K(q^{-1})C(q^{-1})\phi_*^T(k-d)$ is persistently spanning in the sense that "on average positiveness holds" [see (4.49)-(4.51)],
- Both the terms

$$\|[(K(q^{-1})C(q^{-1})-1)v](\cdot)\phi_*(\cdot-d)\|_\infty \qquad (4.52)$$

and

$$\left\|[(K(q^{-1})C(q^{-1})-1)v](\cdot)\frac{\partial\phi_*(\cdot-d)}{\partial\theta}\Big|_{\theta=0}\right\|_\infty \qquad (4.53)$$

are of the order of ϵ^*. ϵ^* is determined by a set of inequalities, which make it possible to linearize, and identify and separate the slow and fast variables, while retaining stability properties.

Under these conditions $\hat{\theta}(k)$ remains close to θ_* and $Y(k)$ close to $Y^*(k)$ provided that the initial conditions are close to θ_* and $Y^*(0)$.

Finally, we comment that the system of inequalities determining ϵ^* does not necessarily have a solution. First we choose a θ_* such as to minimize the tracking error $K(q^{-1})C(q^{-1})-1$ in the frequency range of the input $v(k)$ and to minimize the sensitivity of the regressor ϕ_* with respect to θ_*; then, we calculate, using the linear analysis of Chapter 3 (see, e.g., Theorem 3.1), a bound for ϵ^* and finally verify whether the total stability bounds (4.52) and (4.53) can be satisfied.

In the following chapter more will be said about the practical relevance of these conditions and their interpretations.

5.5 NOTES AND REFERENCES

The course followed through this chapter has been to derive the explicit error systems associated with particular common adaptive algorithms of, identification and control, and then to demonstrate the specific mechanics of appealing to the total stability theorem to establish robust stability conditions. The major goal here is, of course, to identify the individual transfer functions and signals appearing in the linearized adaptive systems for these particular schemes.

The error systems appearing in the ideal analysis, where exact matching is possible, are familiar from global convergence theories of Landau (1976), who first proved the convergence of the output error scheme using hyperstability arguments, and of Goodwin, Ramadge, and Caines (1980), who performed this function for adaptive control.

We reiterate that in these idealized situations asymptotic stability of the error system is identified with convergence of the adaptive scheme. These latter ideas are collectively available in the books of Landau (1979), Goodwin and Sin (1984), and Ljung and Soderstrom (1984).

The error models for nonexact matching in equation error identification are familiar in diverse areas such as adaptive filtering and signal processing since then the equations are linear and may be studied as in Bitmead and Anderson (1980). For nonlinear error systems there are fewer references, although recent works of Kosut and Johnson (1984) and Ortega, Praly, and Landau (1985) are developed along similar lines. The persistence conditions derived in terms of input signals follow the development of Anderson and Johnson (1982a).

Chapter 6

IMPLICATIONS AND INTERPRETATIONS

6.1 INTRODUCTION

This book has displayed a progressively changing tone and emphasis from the general operator-theoretic formulations of Chapter 1 involving broad classes of adaptive systems, to more specialized developments of Chapters 2 and 3 establishing stability conditions for the underlying linear equations, to Chapters 4 and 5 with applications of the previous theory to specific adaptive algorithms, particularly in adaptive control. The approach has been to introduce successively more specific concepts to demonstrate the mechanics of applying the total stability theorem to describe the stability of adaptive systems.

In this chapter, we initially continue this trend further to explore the engineering implications and interpretations of the technical conditions required by the theoretical analysis. In particular, this chapter evolves from presentation of a detailed study of a single adaptive control problem with attention paid to the role of the theory of the book in establishing its adequate performance. Questions of capabilities and limitations of our theory are addressed and discussed with reference to the effects displayed by the example.

Then, with both the theoretical development and expository example complete, the tack changes as a more global view of the system and signal requirements is adopted and their broader implications for adaptive systems is contemplated, especially in terms of feasible engineering constraints necessitated by the theory. To a large extent it is in this component of the chapter that the whole theory is justified in terms of providing design rules to gauge the conditions suitable for the application of adaptive control and to suggest alterations to encourage adaptive control to function well. We certainly shun the view that adaptive control -- or adaptive systems in general -- is fundamentally a black-box technique. Indeed, the characteristics of the problem play a serious role in formulating a robust adaptive system.

A major strength of our theory is that it is not just analytical but, through a careful study of its requirements, can lead to meaningful ideas for the synthesis of "new" adaptation algorithms. The particular modification suggested here is connected with the inclusion of regressor signal filtering prior to adaptation. This prefiltering is shown to be able to assist the satisfaction of the major conditions of robust adaptation but at the price of altered performance. This technique of regressor filtering apparently has been applied by adaptive control practitioners but without being theoretically fully understood. The synergism of theory and practice is cogently displayed by the fact that theory gains direction from such modifications while practical users gain understanding and confidence from available theory.

6.2 ILLUSTRATIVE EXAMPLE

For the following control problem two discrete-time adaptive control solutions will be developed, one direct and one indirect. By its definition the resulting problem will include unmodelled plant dynamics and output measurement noise in these two schemes developed for the ideal case. The insights of this book will be demonstrated, via simulation and analysis, to offer valuable assistance in assuring the robustness of these candidate adaptive controllers.

As this section is relatively long, we shall outline our path briefly here by subsection:

6.2.1 We present in moderate detail a sampled-data reduced-order tracking problem for a particular continuous-time plant and develop some notions of approximate models and suitable controller structures.

6.2.2 Two discrete-time adaptive control algorithms are presented -- one indirect, one direct -- for the problem of Subsection 6.2.1. These correspond to standard approaches similar to those of Chapters 4 and 5.

6.2.3 The step response of the adaptively controlled system is discussed and the destabilizing (drift) effect of additive noise is examined. This represents a situation without a suitable degree of persistence of excitation and demonstrates a lack of robustness.

6.2.4 An acceptable degree of persistence of excitation is achieved by adding a small magnitude sinusoid to the step reference signal of Subsection 6.2.3. The frequency of this reference sinusoid component is varied and the resulting behavior is observed and correlated to the theory of this book.

6.2.1 A Reduced-Order Tracking Problem

The example problem has a model-following configuration. The continuous-time plant with transfer function

$$P(s) = \frac{Y(s)}{U(s)} = \frac{104g}{(s+1)(s+10-j2)(s+10+j2)} \tag{2.1}$$

where g is the dc gain, is to be controlled digitally using a sample period of

$$T = 0.05 \text{ seconds} \tag{2.2}$$

The design is to be based on the assumption that a first-order model with dc gain g provides an adequate approximation of (2.1) over the bandwidth of the input signal u, induced by the reference signal r, during operation. The control objective is predominantly a step reference setpoint tracking task with a desired dc gain of two in the presence of a high frequency, small amplitude sinusoidal output measurement noise. (The high frequency sinusoidal disturbance is simply a convenient "model" for a high frequency bandlimited output disturbance.) In addition, the transients following the infrequent changes in the setpoint level are to follow those of the first-order discrete-time reference model with transfer function

$$M(z) = \frac{X(z)}{R(z)} = \frac{0.2}{z-0.9} \tag{2.3}$$

For $T = 0.05$ as in (2.2), the pole $z = 0.9$ in the reference model corresponds, through $z = e^{sT}$, to $s \approx -2$. This is double the bandwidth of the "dominant" pole of (2.1), i.e., $s = -1$, for which $z = e^{sT}$ yields $z \approx -0.95$.

Since the "unmodelled" poles at $s = -10 \pm j2$ effectively add a double pole only one decade higher than the "dominant" pole at $s = -1$, a compromise plant model for a frequency range extending a bit beyond the dominant pole frequency of 1 rad/sec would have a pole closer than $s = -1$ to the $s-$plane origin. The resulting lower frequency breakpoint with its single pole rolloff represents an attempt to fit better the faster rolloff effects of the unmodelled poles at frequencies above 1 rad/sec. Using such a frequency response fit paradigm, it is easy to see that the actual spectrum of the plant input would determine the parametrization of the least squares fit of a first order approximation of the discretization of $P(s)$. Refer to Ljung (1985) for elaboration of this point. A reasonable compromise model of the plant for design purposes might be

$$\hat{P}(s) = \frac{0.8g}{s+0.8} \tag{2.4}$$

or its zero-order-hold equivalent

$$\hat{P}(z) = \frac{b}{z-a} = \frac{0.04g}{z-0.96} \tag{2.5}$$

Using the controller structure shown in Figure 6.1, i.e.,

$$u(kT) = cr(kT) - dy(kT) \tag{2.6}$$

which is equivalent to the structure used in the similar example of Rohrs et al. (1985), the r to y transfer function, with $P(z)$ replaced by $\hat{P}(z) = b/(z-a)$, can be made to match that of (2.3) with

$$c = 0.2/b \tag{2.7}$$

and

$$d = (a-0.9)/b \tag{2.8}$$

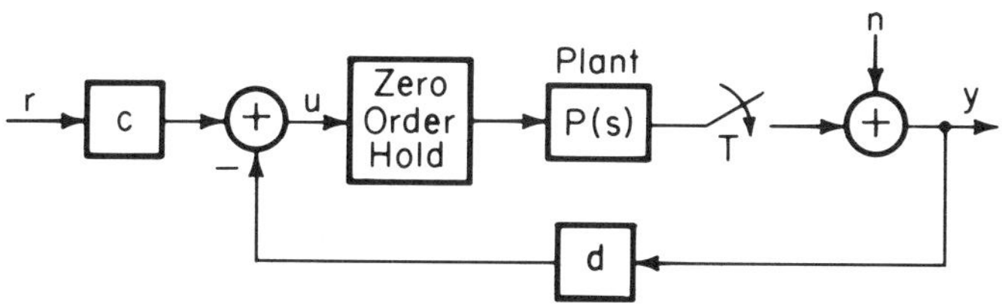

Fig. 6.1 A sampled-data control configuration.

Given $a = 0.96$ and $b = 0.04g$ from (2.5) yields $c = 5/g$ and $d = 1.5/g$. A choice closer to the $s-$plane origin for the pole in (2.4) will increase a in (2.5), and thus decrease b to keep $b/(1-a)$ fixed at g. This will increase both c and d in (2.7) and (2.8).

For a particular d the noise n to output y discrete-time transfer function is

$$\frac{Y(z)}{N(z)} = \frac{1}{1+dP(z)} \tag{2.9}$$

Given the d in (2.8) associated with the approximation $\hat{P}(z)$ of $P(z)$ in (2.5), (2.9) becomes

$$\frac{Y(z)}{N(z)} \approx \frac{z-0.96}{z-0.9} \tag{2.10}$$

Unfortunately, (2.10) does not itself offer a desirable lowpass (or notch) effect. In fact, as d is increased, the highpass character of

(2.10) is undesirably enhanced as long as $1+dP(z)$ remains stable. The transfer function of actual interest, however, is that between the disturbance noise and the actual plant output, i.e., $-dP(z)/(1+dP(z))$. For $P(z)$ approximated by $\hat{P}(z)$ from (2.5), this transfer function is $-0.06/(z-0.9)$. In this transfer function as d is increased the dc gain of this transfer function increases to unity and its pole moves farther to the left along the real axis of the $z-$plane. This indicates that an increasing d, within its stabilizing region, will tend to increase the effect of the high frequency output measurement noise in the actual plant output as well as increase its presence in the measured output y. Thus, the modest bandwidth improvement required by (2.3) represents a compromise between faster transient response decay and less attenuation of the high frequency output measurement noise.

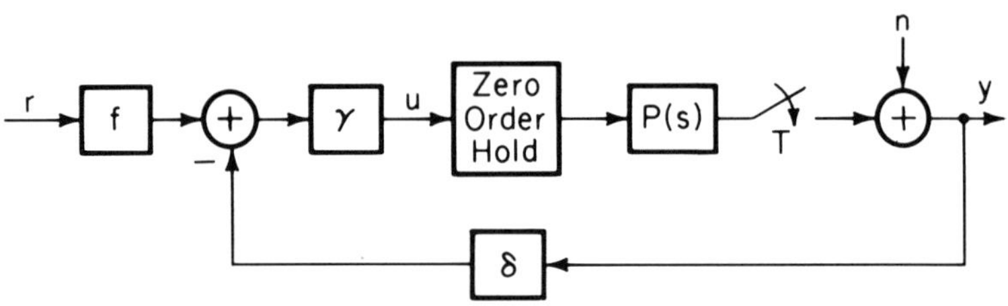

Fig. 6.2 Another sampled-data control configuration.

An alternative control configuration, identical to that used in chapter 5, uses the control law illustrated in Figure 6.2:

$$u(kT) = \gamma[fr(kT)-\delta y(kT)]. \tag{2.11}$$

In (2.11), f is fixed as the numerator of the reference model, i.e., 0.2 in (2.3). Thus, γ and δ can be parametrized from the choice of $\hat{P}(z)$. Using (2.5) in that capacity yields

$$\gamma = 1/b = 25/g \tag{2.12}$$

and

$$\delta = a-0.9 = 0.06 \tag{2.13}$$

Note that $\delta\gamma = 1.5/g$, which equals d for the preceding configuration shown in Figure 6.1. When $\delta\gamma = d$, the noise n to output y transfer function arising from the control structure of Figure 6.2., i.e.,

$$\frac{Y(z)}{N(z)} = \frac{1}{1+(\delta\gamma)P(z)} \tag{2.14}$$

matches that of Figure 6.1 given in (2.9) and approximated in (2.10).

For the control configuration of (2.6) the value of d determines the pole locations of the closed-loop system via a root locus of the zero-order-hold equivalent of the "actual" plant in (2.1). For the control configuration of (2.11) $\delta\gamma$ plays an identical role. The stability range for d (or $\delta\gamma$), is $(-1/g, 16.2/g)$. The d, or $\delta\gamma$, of $1.5/g$ based on a reduced-order plant model moves the closed-loop pole nearest the unit circle closer to the z-plane origin than the pole of the reference model is. In fact, d (or $\delta\gamma$) $= 1.5/g$ nearly corresponds to the real axis breakaway point of the root locus. Thus, the closed-loop plant effectively has a dominant pair of poles near $z = 0.82$. As d (or $\delta\gamma$) increases substantially above $1.5/g$, the dominant pole pair becomes complex and closed-loop system transient behavior will begin to exhibit a significant departure from that of the reference model once the damping ratio of this pair is small. As d (or $\delta\gamma$) increases further, this dominant pair leaves the unit circle. It should also be noted that the zero-order-hold equivalent of $P(s)$ in (2.1), for the "short" sample period of (2.2), has a non-minimum-phase zero that is quite a distance outside the unit circle at $z \approx -2.9$.

6.2.2 Adaptive Solutions

Under the condition that the low frequency dominant pole of the model, e.g., (2.4) of (2.1), is imprecisely locatable before control system activation, an on-line procedure for automatically tuning the controller to achieve acceptable model-following is a prime design candidate. Recall that the optimal solution of the reduced-order tracking problem in the preceding section is dependent on the spectrum of the reference signal, which is expected to vary with time due to the nonstationarity of the distribution of setpoint changes. This suggests the use of an adaptive control procedure, it is hoped exhibiting asymptotic near-optimality, that continues to monitor system performance and appropriately adjust the controller parametrization even after the start-up auto-tuning phase. The use of a small step-size adaptive controller is advisable to provide a desirable long-term averaging of the output measurement noise effects on the average convergent controller parametrization. The smallness of the step-size is limited by the speed of slow (but unpredictable) variations in the desired pole location. Candidate controller parameter adaptation algorithms can be developed for each of the control configurations in Figures 6.1 and 6.2.

For the control law of (2.6) used in Figure 6.1 time indices are added to the controller parameters, as in

$$u(kT) = c(kT)r(kT) - d(kT)y(kT) \tag{2.15}$$

to indicate their time variation due to their adaptation. The time indices of the parameter estimates in (2.15) also tie it to the timing of the adaptive (or update) laws

$$c(kT) = c((k-1)T) + \sigma((k-1)T)\, r((k-1)T)\, v(kT) \tag{2.16}$$

$$d(kT) = d((k-1)T) - \sigma((k-1)T)\, y((k-1)T)\, v(kT), \tag{2.17}$$

where the normalized step-size is

$$\sigma((k-1)T) = \frac{\mu_1}{1+\mu_2(r^2((k-1)T)+y^2((k-1)T))} \tag{2.18}$$

and the prediction error is

$$v(kT) = x(kT)-y(kT)-0.9[x((k-1)T)-y((k-1)T)] \tag{2.19}$$

The signal x in the prediction error is the reference model output

$$x(kT) = 0.9x((k-1)T) + 0.2r((k-1)T) \tag{2.20}$$

Thus, $v(kT)$ could be rewritten without explicit use of an active reference model, i.e., without explicit computation of x, via $v(kT) = 0.2r((k-1)T)-y(kT)+0.9y((k-1)T)$. Recall that y is the sampled output of $P(s)$, due to u, plus a small output measurement noise.

The adapted control law associated with Figure 6.2 is

$$u(kT) = \gamma(kT)[fr(kT) - \delta(kT)y(kT)] \tag{2.21}$$

Compare (2.21) to (2.15). Recall that $f = 0.2$. From Chapter 5, a candidate adaptation scheme for updating γ and δ is provided by

$$\gamma(kT) = \frac{1}{w(kT)} \tag{2.22}$$

$$\delta(kT) = n(kT) \tag{2.23}$$

where

$$w(kT) = w((k-1)T) + \rho((k-1)T)\ u((k-1)T)\ \eta(kT) \tag{2.24}$$

$$n(kT) = n((k-1)T) + \rho((k-1)T)\ y((k-1)T)\ \eta(kT) \tag{2.25}$$

where the prediction error η is

$$\eta(kT) = y(kT) - 0.9y((k-1)T) - w((k-1)T)u((k-1)T) - n((k-1)T)y((k-1)T) \tag{2.26}$$

and the step-size ρ is

$$\rho((k-1)T) = \frac{\epsilon}{1 + \epsilon[u^2((k-1)T) + y^2((k-1)T)]} \tag{2.27}$$

Note that, since $0.2r(kT) = w(kT)u(kT)+n(kT)y(kT)$, from (2.21)-(2.23), the indirect algorithm prediction error η in (2.26) is precisely the negative of the direct algorithm prediction error v in (2.19). Recall the need to avoid division by zero in (2.22), as discussed in Chapter 5.

To support the development of these two candidate adaptive algorithms and to help connect them to the difference equation of principal interest cited in (1.2) of Chapter 2, consider the ideal case.

In the ideal case the plant has an exact first-order model and no output measurement noise is present, that

$$y(kT) = ay((k-1)T) + bu((k-1)T) \tag{2.28}$$

In this ideal case, both of the candidate algorithms fit the generic form

$$\theta(k) = \theta(k-1) - \frac{\epsilon_1 \phi(k-1)\phi^T(k-1)\theta(k-1)}{1 + \epsilon_2 \phi^T(k-1)\phi(k-1)} \tag{2.29}$$

where ϵ_1 and ϵ_2 are both small positive numbers.

To relate the direct algorithm in (2.15)-(2.20) to (2.29), use (2.20) and (2.28) to rewrite the prediction error v in (2.19) as

$$\begin{aligned} v(kT) &= b\{[0.2/b - c((k-1)T)]r((k-1)T) \\ &\quad - [\tfrac{a-0.9}{b} - d(k-1)]y((k-1)T)\} \end{aligned} \tag{2.30}$$

Recall the formulas for c and d in (2.7) and (2.8). Defining

$$\phi(k) = [\ r(kT) \quad -y(kT)\ ]^T \tag{2.31}$$

and

$$\theta(k) = [\ 0.2/b - c(kT) \qquad \frac{a-0.9}{b} - d(kT)\ ]^T \tag{2.32}$$

converts (2.30) to

$$v(kT) = b\phi^T(k-1)\theta(k-1) \tag{2.33}$$

Thus, (2.15)-(2.20) matches (2.29) with $\epsilon_1 = \mu_1 b$ and $\epsilon_2 = \mu_2$.

For the indirect adaptive control scheme of (2.21)-(2.27) the prediction error η in (2.26) can be rewritten in the ideal case as

$$\begin{aligned} \eta(kT) &= [(a-0.9)-n((k-1)T)]y((k-1)T) \\ &\quad + [b-w((k-1)T)]u((k-1)T) \end{aligned} \tag{2.34}$$

The objective of zeroing η supports the acceptance of w as an estimate of b and n as an estimate of a minus the desired reference model pole. Use of the certainty equivalence principle would suggest solving for γ and δ using (2.12) and (2.13) once w and n are converted to the associated a and b. This translation from the estimated w and n to the controller parameters γ and δ is done by (2.22) and (2.23) without intermediate solution for the associated a and b. Defining

$$\phi(k) = [\, u(kT) \;\; y(kT) \,]^T \tag{2.35}$$

and

$$\theta(k) = [\, b - w(kT) \quad (a-0.9) - n(kT) \,]^T \tag{2.36}$$

equates (2.21)-(2.27) to (2.29) with $\epsilon_1 = \epsilon_2 = \epsilon$.

To connect (2.29) to (1.2) of Chapter 2 rewrite (2.29) premultiplied by $\phi^T(k-1)$ as

$$\frac{\phi^T(k-1)\theta(k)}{1+(\epsilon_2-\epsilon_1)\phi^T(k-1)\phi(k-1)} = \frac{\phi^T(k-1)\theta(k-1)}{1+\epsilon_2\phi^T(k-1)\phi(k-1)} \tag{2.37}$$

Using the left side of (2.37) to replace its right side in (2.29) reveals that (2.29) and (1.2) of Chapter 2 are identical if $\epsilon_1 = \epsilon_2$. This is true for the indirect algorithm where $\epsilon_1 = \epsilon_2 = \epsilon$. This equivalence is exact for the direct algorithm only if $\mu_2/\mu_1 = b$ and $\mu_1 b > 0$. As seen from (2.37), as long as $\mu_2 - \mu_1 \max(b)$ remains sufficiently small such that $1+(\epsilon_2-\epsilon_1)\ \phi^T(k-1)\phi(k-1) \approx 1$, where $\max(b)$ is the known bound on the largest magnitude b can acquire, (2.29) is adequately approximated by (1.2) of Chapter 2.

We should point out that the indirect algorithm of (2.21)-(2.27) is a model-following variant of the trajectory-following adaptive controller of Goodwin, et al. (1980) termed "SISO Projection Algorithm I." Similarly the direct algorithm of (2.15)-(2.20) is a variant of "SISO Projection Algorithm II" in Goodwin, et al. (1980). Note that in the ideal case the $H(z)$ of (1.2) of Chapter 2 is transparent as unity, which accounts for the absence of SPR requirements in Goodwin, et al. (1980). But as we saw in Chapter 5 for the indirect algorithm, and will confirm for the direct algorithm of this example later in this chapter, $H(z)$ becomes nonunity in the presence of unmodelled plant dynamics.

6.2.3 Adaptive System Step Response

For the moment we will continue to consider the ideal case. For a step reference input, the plant exactly modelled by (2.28), and no output measurement noise, the algorithms of the preceding section are provably stable, but the parameter error vector θ does not necessarily decay to zero. In fact, assuming that the transients due to unknown initial conditions in the plant have disappeared, any stabilizing controller parametrization maintaining the desired dc gain of 2 will be an equilibrium point due to the provision of zero prediction error. These equilibrium points form a line in parameter space, which we will designate the dc matching line. For the direct scheme

$$\frac{cg}{1 + dg} = M(1) = 2 \tag{2.38}$$

or

$$d = c/2 - 1/g \tag{2.39}$$

defines the dc matching line. For the indirect scheme

$$\frac{f\gamma g}{1 + \delta\gamma g} = M(1) = 2 \tag{2.40}$$

or, using (2.22) and (2.23) and the fact that $f = 0.2$,

$$w = g(0.1 - n) \tag{2.41}$$

defines its dc matching line. (Peek ahead at Figures 6.3 and 6.4, which plot these dc matching lines in addition to providing a graphical summary of experiments described in the following paragraphs. Taking such an advance look at these figures will help in understanding the rather involved remarks of the following paragraphs.) In higher order examples this equilibrium set could be a higher dimension subspace, with its dimension equal to the number of parameters to be estimated minus the rank of the regressor summed outer product.

In Rohrs, et al. (1985), it is demonstrated that with the addition of a small amplitude sinusoidal output measurement noise the parameter estimates will quickly approach the dc matching line and then drift "along" this line toward larger values of $|d|$. The motion along the dc matching line is not a migratory vibration coincident with the dc matching line (Astrom, 1984). In fact, its appearance is a vibration mostly orthogonal to the dc matching line that shows a slow migration within a small tube around the dc matching line. Simulations indicate that the direction taken "along" the line is dependent on the frequency of the disturbance sinusoid. In our example, the simulated drift direction is "up and to the right" along the DC matching line of (2.39) in (c,d) space, i.e., both c and d increasing, for the direct algorithm with $\mu_1 = \mu_2 = 0.01$, $g = 5$, $r(k) = 2$ and low frequency sinusoidal disturbance $\{n(k)\} = 0.04\sin(\omega_d kT)$ with $\omega_d = 4$ and 8 rad/sec. The drift direction is "down and to the left" for high frequency disturbances with $\omega_d = 12$ and 20 rad/sec. The disturbance sinusoid magnitude multiplier of 0.04 was chosen to simulate "1% noise" in the desired 4 unit step output. For the same example, the indirect algorithm drifts "down and to the right" on the dc matching line of (2.41) in (n,w) space, i.e., n increasing and w decreasing, with low frequency disturbances $\omega_d = 4$ and 8. With w corresponding to b, when w drops, b drops, a increases [to keep $b/(1-a)$ equal to the constant g], and c and d in (2.7) and (2.8) both increase. This shows that the observed induced drift direction, in terms of the implicit plant model parameter (a and b) changes, is the same for both the indirect and direct algorithm given low frequency disturbances. Since the observed drift in the $(n,w)-$plane is "up and to the left" along (2.41) for $\omega_d = 12$ and 20 rad/sec, the preceding statement also applies to the drift direction correlation given high frequency disturbances.

For this sampled-data problem, there is a limit on d or $\delta\gamma(=n/w)$ in order for closed-loop stability to be maintained (even for the idealized plant in (2.28)). Clearly (2.39) ultimately crosses any bound on $|d|$ as c increases or decreases without bound, thus implying that this drift, if not halted, could lead to closed-loop system instability. Rewrite (2.41) as

$$n/w \;=\; (0.1/w) - (1/g) \tag{2.42}$$

For w negative, n/w has already crossed the stability boundary of $-1/g(=-0.2)$ and $[1+(\delta\gamma P(z)]^{-1}$ is unstable. Thus, w must remain positive for stability. If w is positive but too small then n/w is too large and instability still results. As w becomes positive and large enough, $[1+\delta\gamma P(z)]^{-1}$ is stable. As w becomes large and positive, n/w approaches the stability boundary of $-1/g$ from the stable side. The relative stability in terms of allowable deviations of (n,w) from the dc matching line is also dropping, which increases the chance of destabilization due to the drift vibrations. Thus, unhalted drift is unacceptable.

This entire book has been based on the premise that ϕ satisfied a persistent excitation condition, e.g., as in (6.14) of Theorem 2.9. In this step response of the adaptive control system, persistent excitation is not present in the ideal case since y converges to $2r$ and u converges to $2r/g$. Our theory here does not lead us to expect local stability with the addition of output measurement noise unless the spectral complexity of r is increased.

6.2.4 Performance with Persistent Excitation

Consider adding a sinusoid to r such that

$$r(kT) \;=\; 2 + 0.1\sin(\omega_r kT) \tag{2.43}$$

For the present example, where two parameters are being estimated, this r is persistently exciting e.g., in (4.10) of Property 3 of Chapter 5. Actually r is complex enough to be persistently exciting for the estimation of three parameters. This indicates that, with the presence of unmodelled dynamics in the plant transfer function, no single fixed choice for the two parameters being estimated in our example will exactly zero the prediction error for a particular ω_r. This fact can be supported through the frequency response fit paradigm. In our problem, to exactly track the reference model's response to r in (2.43), we need to fix simultaneously *three* values on the closed-loop control system Bode plots: the gain at dc and the gain *and* phase shift at ω_r. But we have only two variable closed-loop system parameters, which makes an exact fit impossible in general for a higher than first-order closed-loop system.

Given that the prediction error cannot be zeroed, and thus the update algorithm remains perpetually active, a reasonable question is: Around what average parametrization does the algorithm orbit, if it is locally stable? For the algorithms in (2.16)-(2.17) and (2.24)-(2.25), this is a parametrization that causes the product of the prediction error and the regressor ϕ to have a zero average. Such an average, for a fixed controller parametrization, is readily performed in our example given the periodicity of r. Assume that we have found one θ_* (where $\theta = \theta_* - \hat{\theta}$) that causes this average to be zero. This raises a second question: If the algorithm "reaches" the vicinity of this θ_*, will it remain there? Writing the regressor and prediction error product average as avg$\{\phi e\}$, the answer to this question is yes, if on average

$$\frac{\partial}{\partial\theta}\Delta\theta\,|_{\theta=0} = \frac{\partial}{\partial\theta}[-\phi e]\,|_{\theta=0} = -[e\,\frac{\partial\phi}{\partial\theta} + \phi\,\frac{\partial e}{\partial\theta}]\,|_{\theta=0} < 0 \tag{2.44}$$

In Chapters 4 and 5 e is rewritten, using the tuned system concept, as $H(z)[\phi^T\theta]+e_*$, where $e_* = e\,|_{\theta=0}$. Then, if e_* is suitably small, (2.44) can be approximated by the average of $-\phi H(z)[\phi^T]$ evaluated at $\theta = 0$, i.e., avg$\{\phi_* H(z)[\phi_*^T]\}$. (Refer to section 4.3 for a formal development of this "approximation".) This view heuristically explains the key technical result of this book, which is that the positive definiteness of the average of $\phi_* H(z)[\phi_*^T]$ guarantees local stability if e_* and the algorithm step-size are small enough (and certain other technical conditions are satisfied). The preceding chapters are devoted to establishing the requisite technical conditions, which will be discussed in the next section of this chapter, for proving this central result.

To illuminate this result, we performed simulations of the two adaptive controllers when applied to (2.1) for different ω_r in the reference signal of (2.43) with ω_d of the output disturbance fixed at 20 rad/sec. The magnitude of the reference sinusoid is chosen as somewhat larger than that of the disturbance sinusoid, since we expect that this will be needed to counterbalance the drift motion in the absence of the reference sinusoid. Since the magnitude of the reference sinusoid is still quite small relative to the magnitude of the step component of the reference signal, one would expect the "tuned" setting, selected as the average convergence point, to be near the dc matching line. In fact, we chose to initialize our simulations with controller parametrizations on the dc matching line and with regressor and plant initial conditions as those of steady-state dc matching in the absence of the reference and measurement noise sinusoids. Our purpose was to observe whether or not the drift of the adaptive step response in the presence of noise is "halted" about a stabilizing controller parametrization with the addition of the sinusoid to the reference signal.

A set of such experiments with the direct algorithm with $\mu_1 = \mu_2 = 0.01$ allowed us to break up the dc matching line (for $g = 5$) into three contiguous regions, within which the drift is halted for three particular choices of ω_r (= 4, 6, and 8 rad/sec). We initialized the adaptive controller settings from four different approximate model $\hat{P}(z)$ parametrizations with dc gain 5, i.e., (a,b) = (0.9, 0.5), (0.975, 0.125), (0.986, 0.071), and (0.991, 0.045). Using (2.7) and (2.8) converts these (a,b) pairs to (c,d) = (0.4, 0), (1.6, 0.6), (2.8, 1.2), and (4.4, 2). We initialized the controller settings at these 4 endpoints of three contiguous segments on the dc matching line in (c,d) space and observed the direction of the drift along the dc matching line. We assumed that drift was halted near the dc matching line between two of these 4 endpoints if the directions of the drift, for a particular ω_r, at the two endpoints of that particular dc matching line segment were both into the segment. This approach is supported by the view that at some point on the dc matching line an "average force" exerted by the reference sinusoid equals the negative of the "average force" exerted by the disturbance sinusoid. We have already noted our observance of the drift-causing force of the disturbance sinusoid in the absence of any reference signal component beyond the step. If the added reference sinusoid is to halt this drift, as expected, this force diagram seems an apt image. Our observations in this set of experiments lead to the added generalization that the "net" force contributed by the added reference sinusoid along the dc matching line is toward a particular finite-valued point. As long as the frequency of the added reference sinusoid is appropriately restricted, this point to which it "points" is within the stabilizing range of d. Over the stabilizing range of d along the dc matching line, the "disturbance force" always points in the same direction: down and to the left. Thus, the expectation is that for the high frequency disturbance the ultimate local stability region should be a bit below that that would arise from the step plus sinusoid reference without any disturbance.

Refer to Figure 6.3 for a summary of experimental results. Essentially, as the reference sinusoid's frequency is increased, the pole location of the implicit first-order plant approximation corresponding to the neighborhood of local stability moves closer to $z = 1$. Recall the frequency fit paradigm mentioned previously. This paradigm suggests that the observation above has the tendency expected. Note that the frequencies tested were all above that of the actual low frequency pole of the plant. To match adequately the more severe high frequency magnitude and phase shift drops of the

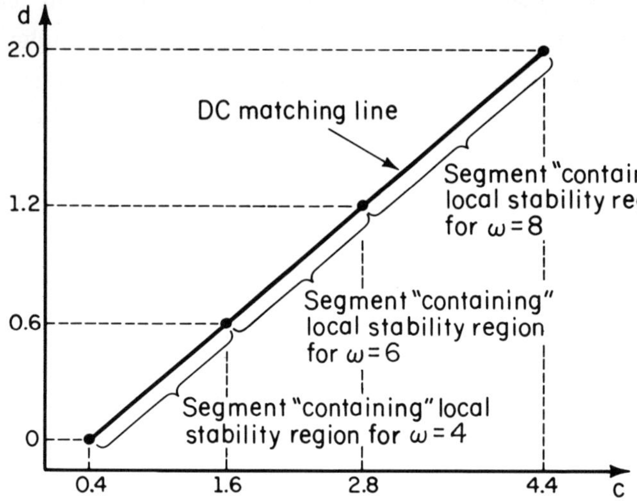

Fig. 6.3 Direct adaptive control dc matching line.

actual third order plant relative to a first order model with the same low frequency pole, the first order model will better approximate the third order plant by placing its single pole at a lower frequency than the lowest frequency pole of the third order plant. As expected, a similar corrective tendency was noted for a fixed ω_r (above the low frequency breakpoint of the plant) as the magnitude of the reference sinusoid increased. As noted earlier, this effect leads to larger d and therefore, above a certain value, to decreasing relative stability as the destabilizing value of positive d is approached. This suggests that if the reference frequency is too high that the point to which the algorithm is driven has a destabilizing d coordinate. Indeed, for $\omega_r = 10$ rad/sec, the drift direction was always to increasing values of d (and c), for every initialization tried within the stabilizing region of the dc matching line. In other words, when the reference sinusoid's frequency is 10 rad/sec, the direct adaptive control system no longer appeared to exhibit local stability.

For $\omega_r = 6$ and 8 rad/sec the direct algorithm was simulated for enough iterations to observe effective drift cessation and the "convergent" local stability behavior. For $\omega_r = 6$ rad/sec, the convergent (c,d) orbits around (1.84, 0.72), which corresponds to an (a,b) of (0.978, 0.1087). For $\omega_r = 8$ rad/sec, the convergent (c,d) orbits about (3.9,1.75), which corresponds to an (a,b) of (0.99, 0.051). To observe the effect, if any, of dc gain g, we decreased the plant dc gain g by an order of magnitude to 0.5 and observed the location of effective drift cessation. A decrease in plant dc gain g without any changes in the approximation of the pole location would all be absorbed by a decrease in b. A decrease of g from 5 to 0.5 suggests that c and d be scaled by a factor of 10, as indicated by (2.7) and (2.8). Scaling the "average convergent" value of (c,d) for $g = 5$ and $\omega_r = 6$ by 10, which yields (18.4, 7.2), provides a reasonable approximation of the "average convergent" (c,d) of (19.4, 7.7), measured for $g=0.5$ and $\omega_r = 6$, since its associated (a,b) of about (0.979, 0.103) corresponds quite well to the implicit (a,b) when $g = 5$.

The indirect algorithm was tested in a similar manner with $\epsilon = 0.001$ and $g = 5$. The initializations of the estimated parameters n and w tested corresponded to solutions of $w = b$ and $n = a - 0.9$ for the (a,b) pairs that generated, respectively, the (c,d) test points above for the direct algorithm, i.e., (n,w) = (0, 0.5), (0.075, 0.125), (0.086, 0.07), and (0.091, 0.045). Interestingly enough, the segments in Figure 6.4 "containing" the local stability regions for ω_r = 4,6, and 8 correspond to those of Figure 6.3 in terms of the implicit (a,b) of the segment endpoints. In fact, for $\omega_r = 6$, (n,w) settled at approximately (0.078, 0.11) and for $\omega_r = 8$ at approximately (0.0887, 0.0565). From (2.12), (2.13), (2.22), and (2.23), $b = w$ and $a = n + 0.9$. Thus, the implicit (a,b) pairs, respectively, for $\omega_r = 6$ and 8, are (0.978, 0.11) and (0.9887, 0.0565). These implicit (a,b) pairs for the indirect scheme are quite close to those measured for the direct scheme under similar

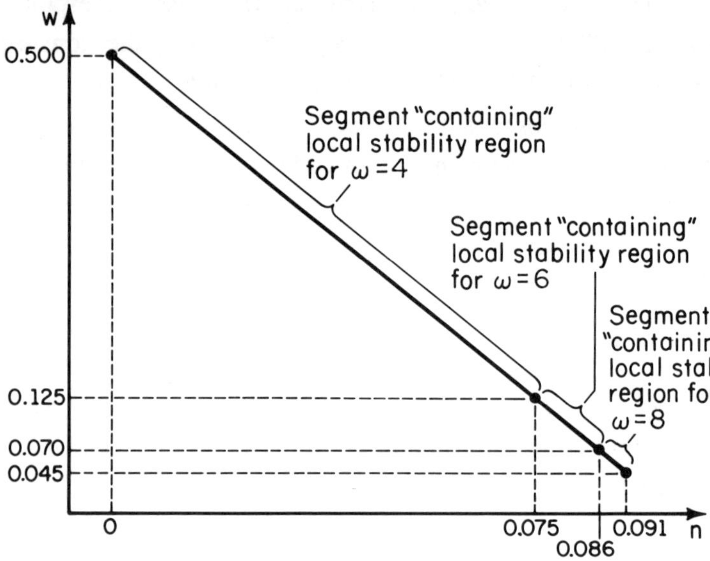

Fig. 6.4 Indirect adaptive control dc matching line.

conditions, i.e., $g = 5$ and $\omega_r = 6$ and 8, cited in the previous paragraph. When g was lowered to 0.5, for $\omega_r = 6$ the convergent (n,w) was measured as (0.138, 0.0727) for an implied (a,b) of (0.9727, 0.138). Note that dropping the dc gain changes the implicit (a,b) in this indirect case slightly more than in the direct case. This can be traced (at least partially) to the fact that the regressor $[r \; -y]^T$ in the direct case is unaffected by a plant dc gain change, while the indirect case regressor $[u \; y]^T$ *is* substantially influenced. This modest difference between direct and indirect adaptive controller robustness properties is indicative of a variety of subtle differences that could be traced to their different definitions of regressor ϕ [and prediction error transfer function modifier $H(z)$].

We would now like to connect the preceding observations of simulated behavior to the analytical insights of this book. We will only compute the "indicators" associated with the direct scheme. (The pertinent formulas for the computation of these indicators for the indirect scheme will be provided parenthetically.) First we need to extract the tuned error system description. To construct the tuned system from Figure 6.1 describing the direct scheme, recognize that the prediction error v used in (2.16) and (2.17) is the difference of the plant and model outputs filtered through $1-0.9z^{-1}$, i.e., the denominator of the reference model. This results in Figure 6.5a, where $\hat{c}$ and $\hat{d}$ are the current parameter estimates. Replacing $\hat{c}$ and $\hat{d}$ by tuned values c_* and d_* in Figure 6.5b and then subtracting $\phi^T\theta$ $[=\phi^T(\theta_*-\hat{\theta})]$ at the left summing junction leaves u unaltered. When $\theta = 0$, the resulting output of Figure 6.5b is

$$e_* = (1-0.9z^{-1})\left\{\left[\frac{0.2z^{-1}}{1-0.9z^{-1}} - \frac{c_*P(z)}{1+d_*P(z)}\right]r + \left[\frac{1}{1+d_*P(z)}\right]n\right\} \qquad 2.45$$

where $P(z)$ is the zero-order-hold equivalent of $P(s)$ in (2.1). Thus, by block diagram analysis,

$$\begin{aligned} e &= H(z)[\phi^T\theta] + e_* \\ &= (1-0.9z^{-1})\left[\frac{P(z)}{1+d_*P(z)}\right][\phi^T\theta] + e_* \end{aligned} \qquad (2.46)$$

which reveals that

$$H(z) = \frac{(1-0.9z^{-1})P(z)}{1+d_*P(z))} \qquad (2.47)$$

and

$$\frac{\partial e}{\partial \theta}\Big|_{\theta=0} \approx H(z)[\phi_*^T] \qquad (2.48)$$

where $\phi_* = \phi|_{\theta=0}$ with the noise n removed, as will be detailed in the next paragraph. Note that the $H(z)$ in (2.47) is essentially a scaled version of the tuned transfer function divided by the desired

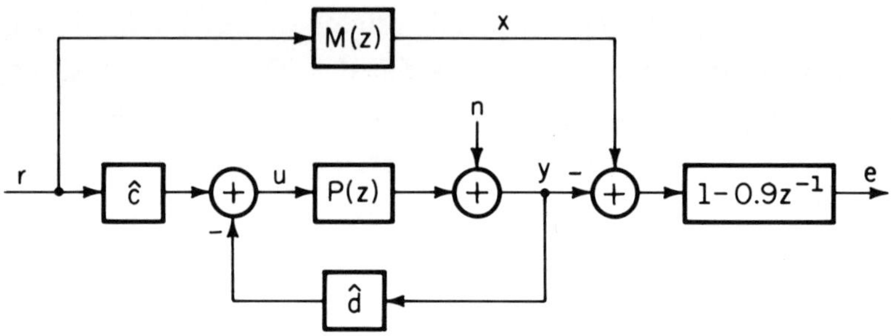

(a) Direct adaptive control system.

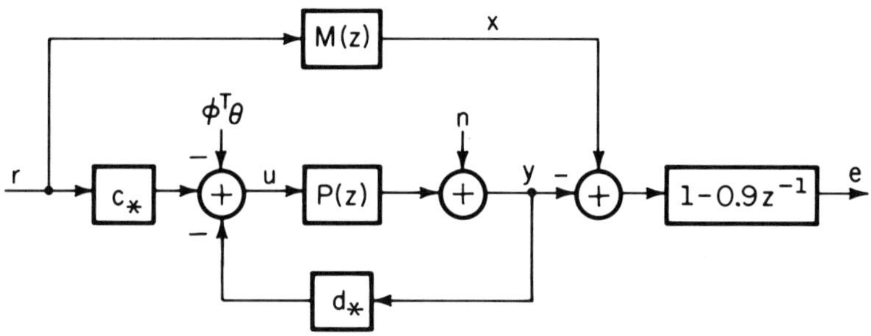

(b) Tuned system extraction from (a).

Fig. 6.5 Tuned error sysem construction.

reference model transfer function. [A similar technique for extracting the tuned error system for the indirect scheme based on Figure 6.2 would result in $H(z)=(1-0.9z^{-1})\gamma_* P(z)/(1+\gamma_*\delta_* P(z))$. Recall that $\eta = -\nu$.]

Next we need to examine the regressor ϕ from (2.31). From block diagram analysis

$$y = \frac{c_* P(z)}{1+d_* P(z)}\, r - \frac{P(z)}{1+d_* P(z)}\, \phi^T\theta + \frac{1}{1+d_* P(z)}\, n \qquad (2.49)$$

assuming that the plant has zero initial conditions (as was essentially enforced in the local stability test simulations) or after enough time that the effects of nonzero initial conditions have disappeared. Considering the effect of the output measurement noise n to be negligible yields

$$\phi \approx \left(r \quad \frac{P(z)}{1+d_*P(z)}\,\phi^T\theta - \frac{c_*P(z)}{1+d_*P(z)}\,r \right)^T \tag{2.50}$$

Evaluating this ϕ at $\theta = 0$ yields

$$\phi_* = \phi|_{\theta=0} \approx \left(r \quad \frac{-c_*P(z)}{1+d_*P(z)}\,r \right)^T \tag{2.51}$$

From (2.50)

$$\frac{\partial\phi}{\partial\theta} \approx \left(0 \quad \frac{P(z)}{1+d_*P(z)}\,\phi_* \right)^T \approx \begin{bmatrix} 0 & 0 \\ \frac{P(z)}{1+d_*P(z)}\,r & \frac{-c_*P^2(z)}{(1+d_*P(z))^2}\,r \end{bmatrix}$$

$$\approx \frac{\partial\phi_*}{\partial\theta_*}\Big|_{\theta_*=[c_*\ d_*]^T} \tag{2.52}$$

(Recall from (2.35) that for the indirect scheme $\phi = [u\ y]^T$ so $\phi_* \approx \left[\frac{0.2\gamma_*}{1+\gamma_*\delta_*P(z)}\,r \quad \frac{0.2\gamma_*P(z)}{1+\gamma_*\delta_*P(z)}\,r \right]^T$.)

Since r is periodic, we are now in a position to compute c_* and d_*, for a particular ω_r in r, that zero the average of the product of ϕ_* in (2.51) and e_* in (2.45) (with the noise n removed). As conjectured above, these candidate equilibrium points closely fit the dc matching line up through $\omega_r = 8$, as shown in Figures 6.6 and 6.7. Figure 6.6 shows the dc matching line and the locus of $\text{avg}\{\phi_*e_*\} = 0$ candidate equilibrium points in the (c,d) plane as ω_r varies. Figure 6.7 plots the d_* causing $\text{avg}\{\phi_*e_*\} = 0$ as a function of ω_r. Figure 6.7 can be used with Figure 6.6 to note the compression of lower frequencies in the reference sinusoid at the lower left of the dc matching line.

The comments following (2.44) imply that if the minimum eigenvalue of the average of $\phi_*H(z)[\phi_*^T]$ is sufficiently positive and e_* is sufficiently small that local stability should result. The

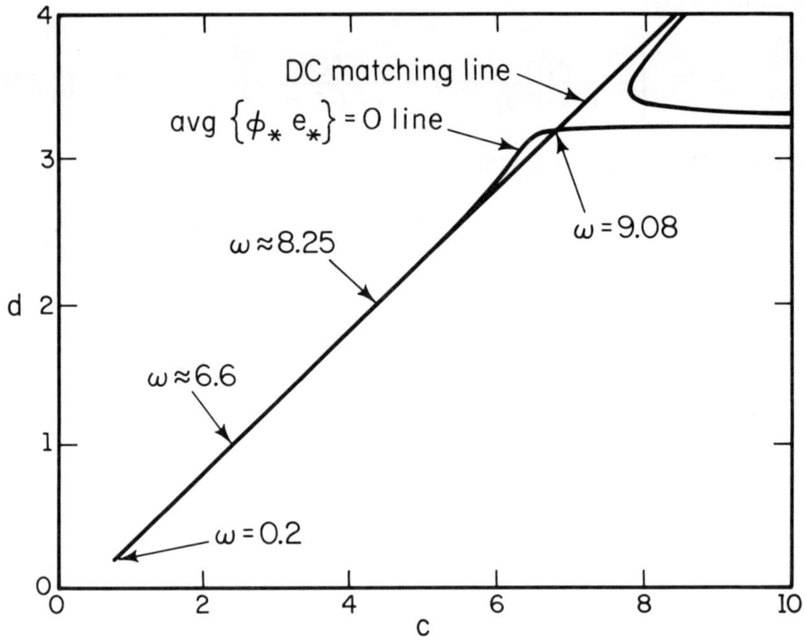

Fig. 6.6 Direct adaptive control dc matching and avg($\phi_* e_*$) = 0 lines.

eigenvalues of the average of

$$\phi_* H(z)[\phi_*^T] \approx \begin{bmatrix} rH(z)r & -r\left(\dfrac{c_* P(z)H(z)}{1+d_* P(z)}r\right) \\ -\dfrac{c_* P(z)}{1+d_* P(z)}r & \left(\dfrac{c_* P(z)}{1+d_* P(z)}r\right)\left(\dfrac{c_* P(z)H(z)}{1+d_* P(z)}r\right) \end{bmatrix} \tag{2.53}$$

can be computed for any (c,d) designated as a tuned setting. Since any (c,d) near the dc matching line of (2.39) will (presumably) lead to small e_*, we could evaluate (2.53) along that locus. For $\omega_r = 2, 4, 6, 8$, and 10 and $g = 5$, this results in Figure 6.8 where the minimum eigenvalue of $\phi_* H(z)[\phi_*^T]$ is plotted as d_* varies along the dc matching line. Note that over the range of d tested in Figures 6.3

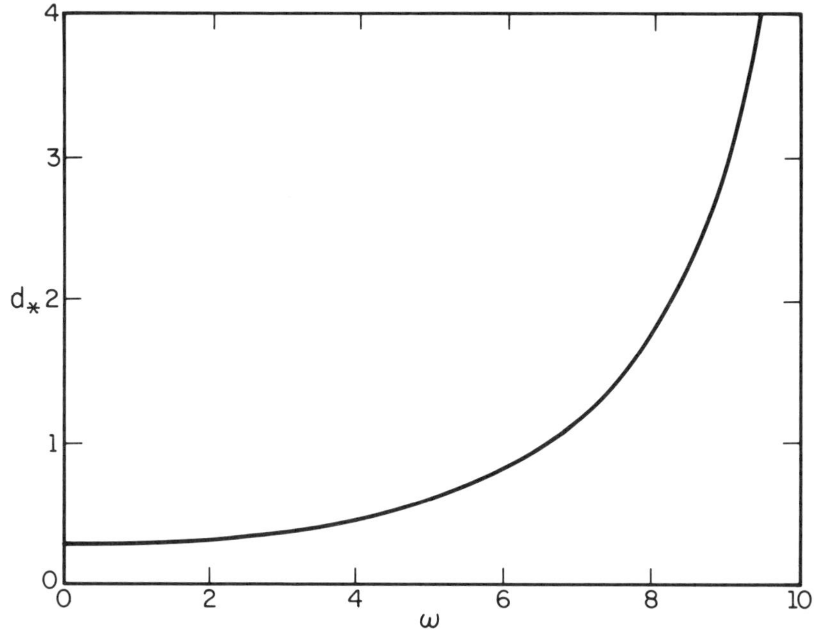

Fig. 6.7 Direct adaptive control avg$\{\phi_* e_*\} = 0$ locus of feedback parameter vs. reference excitation frequency.

and 6.4, i.e., zero to 2, the minimum eigenvalue of $\phi_* H(z)[\phi_*^T]$ is positive for $\omega_r = 4$, 6, and 8, which agrees with the local stability behavior observed in simulation. Note that for all stabilizing positive d, when $\omega_r = 10$ the minimum eigenvalue of $\phi_* H(z)[\phi_*^T]$ is always negative, which corroborates the unhalted drift observed in simulation.

We will return to this example in the next section where we discuss, in more detail, the technical conditions for local stability. Our intent is not to belabor this particular example, but rather to use it as a vehicle in interpreting the implications for adaptive control.

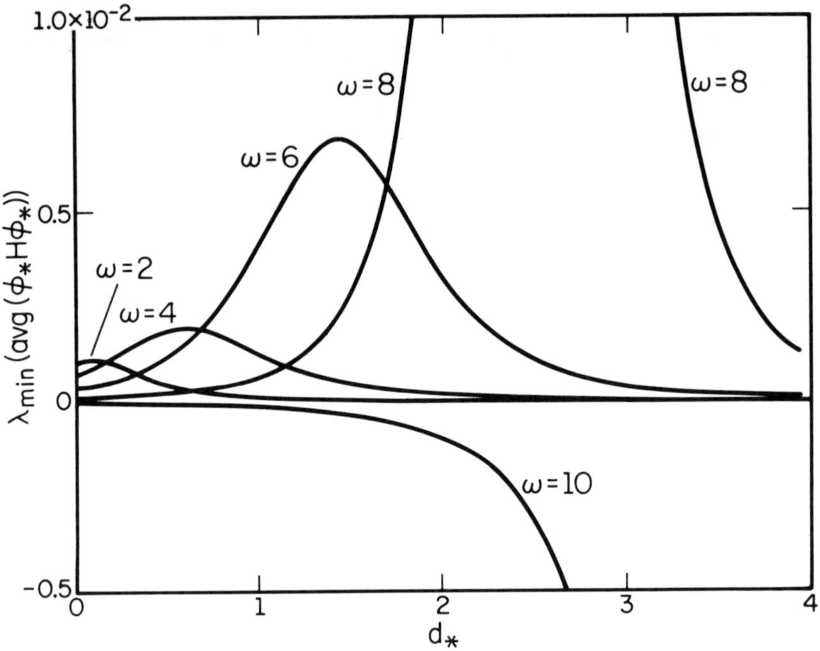

Fig. 6.8 Direct adaptive control signal positivity test in stabilizing segment of dc matching line.

6.3 IMPLICATIONS FOR ADAPTIVE CONTROL ENGINEERING

Refer to Figure 6.9 for a map of the theoretical path constructed in the preceding chapters. This map summarizes the key steps of the adaptive control system stability analysis detailed in this book with specific labelling of the intermediate results, analytical tools used, and assumptions made. Chapter 1 introduces the use of linearization and total stability to analyze the stability properties of

$$\theta(k+1) = \theta(k) - \epsilon\phi(k)\{H(z)[\phi^T(k)\theta(k+1)] + e_*(k)\} \quad (3.1)$$

In this section we consider the generic implications for adaptive control engineering of our theory, as presented in Chapters 2 through 5, regarding the stability of (3.1). Engineering implications will be

Issue/Result	Analytical Tool	Assumptions
Adaptive control		
	Tuned error system construction	(i) Stabilizing θ_* selection
Tuned error system		
	Linearization	(ii) Small $\theta_* - \hat{\theta}(0)$ and small $\phi_*(0) - \phi(0)$ (iii) Well-behaved $\frac{\partial \phi_*}{\partial \theta_*}$ and $\frac{\partial H(z)}{\partial \theta_*}$
Local linearized error system		
	Averaging (or small gain) theory	(iv) Small step-size ϵ (v) Signal positivity ($\phi_* H(z)[\phi_*^T] > 0$)
Local EAS homogeneous error system		
	Total stability	(vi) Small error system disturbance e_*
Bounded error signals		

Fig. 6.9 Theory map.

derived from an interpretation of the assumptions made in our analysis. In several instances it will prove useful to refer to details from the example of the preceding section.

In reading Figure 6.9 and the following discussion of the engineering implications of our theoretical assumptions, it is useful to recall the meaning of the various symbols: θ_*, ϕ_*, θ, $H(z)$, ϵ, and e_*. The parameter error vector θ being updated via (3.1) is the difference of a tuned setting θ_* and the estimated parameter vector $\hat{\theta}$, i.e., $\theta = \theta_* - \hat{\theta}$. It is about this tuned setting that we choose to test local stability. The estimated parameter vector $\hat{\theta}$ can be either a vector of estimates of the plant parameters, which are used to solve for the controller parameters, as in (2.21)-(2.27), or a vector of the directly adapted controller parameters themselves, as in (2.15)-(2.20). If the controller were parametrized using θ_*, the regressor (or partial state) of this tuned control system would be ϕ_*. The actual regressor of the adaptive control system is denoted as ϕ. The transfer function operator $H(z)$ is that through the tuned system from $\phi^T\theta$ to the prediction error used in the updates. Refer to Figure 6.5. This $H(z)$ was denoted H_{ev} in Chapters 4 and 5. In the formulations of Section 6.2, $H(z)$ is a scaled version of the transfer function of the tuned control system divided by the desired (or model reference) transfer function. When the control system can be tuned to exactly match the desired transfer function, $H(z) = 1$. If the plant includes unmodelled dynamics or if output measurement noise is present, the tuned control setting will not zero the tracking error and the error system is subject to a forcing function e_*. The remaining symbol to be defined is the adaptive algorithm update step-size (or scaling factor) ϵ.

As a technical aside, note that the tuned system, about which we wish to test local stability, actually includes a selection of θ_*, ϕ_*, and e_*. The ϕ_* and e_* selected need not, in general, correspond to those associated with θ_* via (4.1) in Chapter 1, as implied in the preceding paragraph. Such a choice is, however, quite natural. It also allows us to regard the local stability problem as about just θ_*.

Given this refresher on the terminology (and analysis objective) of the error system formulation of an adaptive control system, we now proceed directly into an examination of the assumptions indicated in Figure 6.9.

6.3.1 Stabilizing θ_*

The most obvious assumption regarding θ_* selection is that it lead to a stabilizing controller of the *actual* plant. Stabilizing means that the tuned control system regressor ϕ_* (and, thus, the tuned prediction error e_* for finite θ_*) is bounded for all possible reference signals to the control system.

The most basic implication of this assumption is that for the controller structure, within which the parameters are being adapted, some parameter setting must exist that stabilizes the actual plant. This coincides with a major view of the objective of adaptive control as the automatic adjustment of controller parameters to some effectively time-invariant, desirable (i.e., at least stabilizing and possibly more) setting. This suggests that certain problems are simply beyond the reach of adaptive controllers utilizing small step-sizes to insure this effective parameter estimate time-invariance at "convergence." For example, a plant with two unstable real poles and a real zero between these two poles cannot be stabilized by fixed versions of either of the controllers in Section 6.2 illustrated in Figures 6.1 and 6.2. A simple root locus argument supports this statement. Recall that the use of a small step-size ϵ is predicated on the assumption that cautious approach to an essentially fixed setting will prove satisfactory. Thus, neither of the algorithms of the preceding section using a small step-size can be expected miraculously to stabilize such a system. Here the problem is not in the adaptation mechanism itself but in the controller structure being recursively parametrized. Other more subtle control problems are also beyond rescue by small step-size adaptive control algorithms. Remember this: *Adaptive control is not a universal panacea.*

6.3.2 Accurate Initialization

This assumption requires that the initial parameter vector estimate $\hat{\theta}(0)$ be sufficiently close to θ_* and the initial regressor $\phi(0)$ to be sufficiently close to the tuned regressor $\phi_*(0)$ for the particular plant initial conditions. This assumption suggests the need for adequately accurate adaptive controller initialization.

This initialization assumption is an artifact of the local stability question we are addressing: If $\hat{\theta}$ reaches the vicinity of θ_* will it be retained close to θ_*? We are not attempting to discover whether or not $\hat{\theta}$ will close in on θ_* and ϕ_* from a substantial initial distance away, though apparently such an issue can be investigated via extensions of the linearization and total stability theory approach taken in this book. Therefore, we will not draw any significant implications from this assumption.

We will simply cite one interesting interaction of this accurate initialization requirement with step-size ϵ selection. The size of the local stability region (or ball) about a particular $\theta_*(0)$ and $\phi_*(0)$ is directly proportional to the linearized error system exponential convergence rate. Refer to Chapter 1. Also recall that an increase in an already small (positive) ϵ leads to an increase in the convergence rate of the homogeneous version of (3.1). This has to be balanced against the fact that an increase in ϵ also increases the perturbation ϵe_* to θ in (3.1). Thus, increasing ϵ increases the size of the ball about the chosen initialization within which local stability is retained *and* increases the expected excursions in θ (and thus $\tilde{\phi} = \phi_* - \phi$). How this balance will manifest itself in (as yet undeveloped) theoretical guarantees of algorithm attraction to these regions of local stability is an important concern with potential practical significance.

6.3.3 Tuned Controller Insensitivity

With the focus on stable adaptive control system behavior *in the vicinity of the tuned setting,* it is critical that, as the tuned setting is varied within this local region, the character of ϕ_* and $H(z)$ do not vary significantly with modest variations in θ_*. Theoretically, without this assumption the validity of the linearization step collapses.

This assumption implies that the performance of the controller to be adapted should not degrade too rapidly with slight variation in its parametrization. This requirement can be used, in some situations, to justify less complex controllers as better candidates for adaptation (Goodwin and Ramadge, 1979). This assumption does suggest an avoidance of compensation structures that rely on cancellation of singularities very near the stability boundary. It absolutely prohibits the application of plant-numerator-cancelling controllers, which have been quite popular schemes for adaptive implementation, to plants with unstable inverses. This restriction is reflected in the adaptive control arena by the well-recognized minimum-phase assumption of adaptive model-following and minimum variance output regulation (with no cost for control effort). More subtle situations exhibiting excessive sensitivity to controller parametrization are also to be avoided as candidates for small step-size adaptive controller application.

This sensitivity to variations of ϕ_* and $H(z)$ with variations in θ_* also exposes a distinction between indirect and direct adaptive control. In direct adaptive control this sensitivity arises as that of tuned system operators and signals to the directly adapted controller parameters. In indirect adaptive control the tuned system description depends on the translation from the estimated parameters, possibly the plant parameters themselves, to the controller parameters. Thus, the sensitivity of the control design calculations strongly influences the sensitivity of ϕ_* and $H(z)$ to changes in θ_*. Limited sensitivity certainly implies that local continuity about θ_* of the controller parameters as a function of the plant parameters is required as an additional condition to achieve reasonable sensitivity in indirect adaptive control.

As an example of the recommended avoidance of such "design-induced" sensitivity in indirect adaptive control, recall the necessity of avoidance of division by a (near) zero w in (2.22). Actually, as w (or the b being sought) approaches zero, the division in (2.22) certainly enhances the change in the controller parameter γ as w varies by a fixed amount. Thus, the sensitivity of $H(z)$ is increased as b decreases or, equivalently for a fixed g, a approaches unity. Since more rapid sampling, or a smaller T in (2.2), will cause a to approach unity, this advises against selection of an excessively fast sample rate in the indirect adaptive controller of the preceding section.

This connection of the tuned system sensitivity to control system specification constraints can also be imagined in other indirect approaches to adaptive control. One guideline that has emerged for robust indirect adaptive pole placement is the avoidance of estimated parameter vector values corresponding to an estimated plant model transfer function with a pole-zero cancellation. Such a cancellation leads to an infinite control gain request for an arbitrary pole placement objective. This will certainly create an enhanced sensitivity of the tuned system operators and signals about parameter estimate vector settings near those exhibiting plant model pole-zero cancellations. The practical implications in terms of a priori plant model parameter knowledge sufficient to constrain the parameter estimate search to a region excluding nonminimal models is of practical concern.

6.3.4 Small Step-Size

This assumption is a key ingredient in the use of averaging (and small gain) theory on the linearized error system. A key theoretical use of small ϵ is to decouple the time-scales of ϕ and θ variations so $H(z)[\phi^T\theta]$ can be approximated by $H(z)[\phi^T]\theta$ for lowpass $H(z)$. In our analysis, it has been repeatedly stated that step-size smallness does not imply a vanishingly small value. For example, refer to the wording of Theorem 3.1. In fact, we point out that a nonzero bound on how small is adequate can be established. Such bounds are nontrivial to derive and, therefore, typically quite conservative.

The effects of the choice of ϵ within this allowable range deserve attention. One important result is that in this range the convergence rate of the adaptive system is directly proportional to ϵ. Recall that the ultimate objective of our stability analysis, as stated at the bottom of Figure 6.1, is to demonstrate that the adapted parametrization stays close to that of the (effectively time-invariant) tuned control setting and, thus, the performance of the adaptive control system is close to that of the tuned system. Obviously, the larger ϵ becomes, the more the parameter estimates in (3.1) react to the "unremovable" e_* impinging on the error system. Thus, in (3.1), even once the parameter estimates match the tuned setting, a smaller ϵ is needed to attain improved closeness to θ_* at the next iteration. Over several iterations, smaller ϵ corresponds to averaging the effects of e_* on θ over a longer time window.

If the disturbances to the error system are due to the behavior of unmodelled plant modes or output measurement noise, they are to be ignored once we near the tuned setting. This suggests that ϵ should be chosen well below the bound suggested for applicability of averaging (or small gain) theory for improved asymptotic closeness to the fixed θ_*. Conversely, if the disturbances to the error system are due primarily to the time-variations of the tuned setting, then ϵ should be chosen close to this upper bound to achieve as accurate tracking as possible of the changing θ_*. This compromise in the selection of the adaptive step-size is a well recognized feature of adaptive systems.

6.3.5 Signal Positivity

It is this condition on the positivity of

$$\sum_{i=k}^{k+J-1} \phi_*(i)H(z)[\phi_*^T(i)] > 0 \tag{3.2}$$

for all k, that implies a geometrical contraction property for the homogeneous, linearized error system. When ϕ_* is periodic and J is its period, (3.2) reduces to the $\text{avg}\{\phi_* H(z)\phi_*^T\} > 0$ condition arising from (2.44). Satisfaction of this condition can be viewed as primarily a function of appropriate excitation of the control system by the reference input.

A useful interpretation of this requirement arises from a discrete spectrum interpretation of ϕ_* as in Section 3.5.3. Separate ϕ_* into "positive-weighted" components at frequencies for which $\text{Re}[H(e^{j\omega T})]$ is positive, and "negative-weighted" components at those frequencies for which $\text{Re}[H(e^{j\omega T})]<0$. Simply put, (3.2) requires that the "Re[H]−weighted energy" of the positive-weighted components of ϕ_* must exceed the Re[H]−weighted energy of the negative-weighted components of ϕ_*. Recall that ϕ_* is the regressor of the fixed, tuned control system. Thus, for this time-invariant control system the frequencies in the reference input r will appear in the plant input u, the plant output y, and the reference model output x. These measurable signals $r, u, y,$ and x are all that is available to use in forming a regressor for the tuned control system. Refer to (2.31) and (2.35). Thus, the spectrum of the reference signal strongly influences the spectrum of ϕ_*, which is one key to the satisfaction of (3.2).

The most obvious implication of this interpretation of (3.2) is not to allow the reference signal to have significant energy in its negative-weighted components. This implies that the dominant energy band of $\{r\}$ should be contained within the frequency band for which $H(z)$ has a significant gain and a phase shift within $\pm$ 90°. In the examples of Section 6.3.2, $H(z)$ is proportional to the tuned control system transfer function divided by the reference model transfer function. Thus, the character of $H(z)$ is influenced by both the selection of θ_* and the statement of the control objective. If the objective bandwidth included the frequencies of unmodeled modes in the plant, the controller could not be parametrized to compensate for all of the plant modes within the desired flat passband range. Thus, $H(z)$ would not necessarily be appropriately near unity at the significant frequencies in the reference signal. Presumably the frequency range of r would correspond to the bandwidth of the control objective. If the reference signal were not to extend to such high frequencies, there would be little reason to request such a high bandwidth objective. So satisfaction of (3.2) is also dependent on the control designer not requesting that the objective bandwidth be too high relative to the achievable bandwidth with the control structure chosen.

One tendency in satisfying (3.2) might be to restrict the dominating frequency range of the reference signal to well below the achievable bandwidth of the control system, and thus well within the $H(z) \approx 1$ range. This procedure for satisfying (3.2) has its limits. They arise due to the need for the minimum eigenvalue of the left side of (3.2) not to be too small at any k. This need can be seen from the approximation reducing (2.44) to (3.2). Essentially the first term on the right of (2.44) has been neglected in the derivation of (3.2). Since this neglected term is proportional to the nonzero e_*, the minimum eigenvalue of the left of (3.2) cannot be too small. Thus, (3.2) should actually be cast as being greater than some positive σ. Now reconsider pressing the spectrum of the reference signal so far into the range where $H(z) \approx 1$ that it is also predominantly within the bandwidth of the tuned system. Consider the regressor of the direct adaptive control example in (2.31). If r is within the flat passband of the tuned system, then $y_* \approx g_* r$. For the indirect adaptive control example of Section 6.2, ϕ_* from (2.35) would be composed of $y_* \approx g_* r$ and $u_* \approx g_* r/g$, where g_* is the dc gain of the tuned system and g is the dc gain of the plant. Thus, since $H(z) \approx 1$, over the entire range of r, (3.2) will be nearly singular despite the richness of $\{r\}$ over these low frequencies. Thus, its minimum eigenvalue, though positive, could be too small. Such severe low frequency compaction of the spectral content of $\{r\}$ is not recommended.

The undesirable character of the squeezing of the spectrum of $\{r\}$ to within the flat passband of the desired/tuned control system [with $H(z)$ very nearly one over this frequency range] can also be supported by the frequency response fitting paradigm. Imagine attempting to identify a system from behavior due to an input spectrally constrained to within a flat passband in the system response. Attempts to extrapolate the system frequency response curves into the fluctuating portions of the frequency response from this information would be quite sensitive to measurement errors. This undesirable loss of numerical robustness would be suffered in such a case despite the number of spectral lines in a reference signal with such an overly severe bandwidth limitation.

This implies that the dominant portion of the expected lowpass reference signal spectrum should extend beyond the cutoff frequencies of the actual plant and tuned control system. But the reference signal spectrum should not extend too far beyond this achievable bandwidth so as not to put too much energy into the region where $H(z) \neq 1$. This implies that there should be a reasonable width band between the desired bandwidth and the onset of unmodelled effects in the plant. Note that the satisfaction of (3.2) still requires a minimum number of significant spectral lines in this dominant spectrum of $\{r\}$, which is the connection to the now widely accepted definition of persistent excitation (Anderson and Johnson, 1982a; Bitmead, 1984).

6.3.6 Small Disturbances

The disturbances entering the error system are required to remain small. This requirement arises with the use of the total stability theorem. Essentially, the total stability theorem states that a nonlinear system, if not perturbed too far from a homogeneous, EAS linearization, will exhibit a bounded input, bounded state behavior for sufficiently small inputs. The disturbances impinging on the error system are essentially those signals reaching the prediction error driving (3.1) when the adaptive system parametrization is fixed at the tuned setting. The resulting error system disturbance is denoted as e_* in (3.1).

The most immediate interpretation of the requirement that these error system disturbance inputs remain small is that the tuned control system should provide performance adequately close to that desired. From (4.1) in Chapter 1, where $e_* = H_{ew}(\theta_*)w$, it is apparent that e_* is influenced by (i) the character of the external signals in w, such as the reference and disturbance inputs, (ii) the choice of the structure of the control system, which influences the form of H_{ew}, and (iii) the particular tuned controller parametrization via θ_*. See (2.45) for such a description of e_* in the example of Section 6.2. Consequently, in order to try and reduce e_* one must consider seriously the particular design of the control system with regard to the specific exogenous signals. For example, if the control system is expected to be subject to an appreciable constant offset in the output measurement, the control structure, within which the parameters are to be adapted, should possibly include appropriate integral action. As another example, note that normally the nominal plant model used for controller design is based on the dominant open-loop modes of the plant. To reduce adequately e_* it may become necessary to incorporate more of the less dominant open-loop plant modes in the nominal design procedure. The control objective will also play an important role in determining the (satisfactorily) dominant order. This comment is readily supported by an argument based on the overlap of the bandwidth of less dominant open-loop plant modes and the bandwidth of the control objective.

A reasonable question is how small is small enough for e_*. For a partial answer, recall (2.44). The signal positivity condition on $\phi_* H(z)[\phi_*^T]$ alone, as in (3.2), is based on the assumption that e_* is small enough so that the e_* term on the right of (2.44) can be ignored. This indicates the interplay between the magnitude of e_* and the magnitude of the smallest eigenvalue of the positive definite "average" $\phi_* H(z)[\phi_*^T]$. To illustrate this compromise reconsider the direct adaptive control simulations summarized in Figure 6.3. The smallest eigenvalue of $\text{avg}\{\phi_* H(z)[\phi_*^T]\}$ is plotted in Figure 6.8 as evidence that the signal positivity condition is satisfied along the dc matching line for various frequencies of the sinusoid in the reference signal of (2.43). Note that the smallest eigenvalue of $\text{avg}\{\phi_* H(z)[\phi_*^T]\}$ is on the order of the magnitude of the disturbance for a large portion of the stabilizing region of the dc matching line. Since the noise to reference transfer function in (2.9) enhances its highpass character with increasing d_*, as d_* is increased e_* is increased. Also, as d_* is increased the unmodelled modes become more significant in the achieved control system transfer function. A higher d_*, by increasing the achieved bandwidth, is also positioning the reference sinusoid deeper into the passband of $H(z)$, thereby reducing the minimum eigenvalue, as discussed in the preceding subsection. Thus, for larger d_* within the stabilizing range, the potential exists that e_* is large enough so $\text{avg}\{\phi_* H(z)[\phi_*^T] + e_* \frac{\partial \phi_*}{\partial \theta_*}\}$ is no longer positive definite.

That this indeed happens in this example is illustrated by Figure 6.10. Figure 6.10 repeats the plot of the minimum eigenvalue $\text{avg}\{\phi_* H(z)[\phi_*^T]\}$ along the dc matching line with $\omega_r = 8$ from Figure 6.8. On the same coordinates the minimum eigenvalue of $\text{avg}\{\phi_* H(z)[\phi_*^T] + e_* \frac{\partial \phi_*}{\partial \theta_*}\}$ is plotted. (Note that the computation of e_*

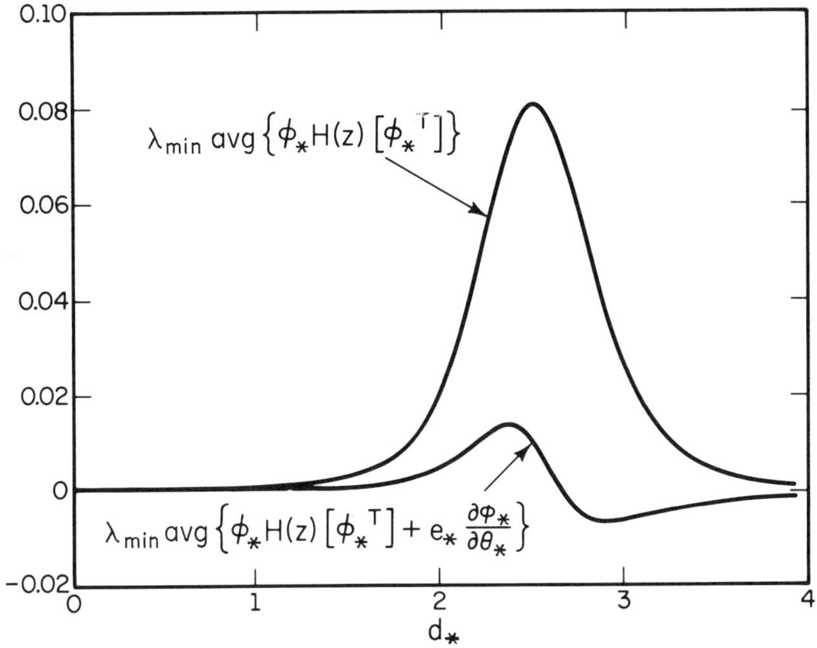

Fig. 6.10 Direct adaptive control: avg$\{\phi_* H(z)[\phi_*^T]\}$ and avg$\{\phi_* H(z)[\phi_*^T] + e_* \partial\phi_*/\partial\theta_*\}$ comparison along dc matching line for $\omega_r = 8$.

for Figure 6.10 excludes the noise sinusoid and therefore represents only an approximation -- albeit, adequate here -- of its "actual" value.) This latter curve crosses zero into negative values with d_* above approximately 2.6. This indicates that, if the controller is initialized on the dc matching line with a d_* above 2.6, local stability will not be attained. Our theory does not say anything about the more global behavior of the algorithm. Note in Figure 6.10 that, for some stabilizing parametrizations along the dc matching line, e_* is small enough so that satisfaction of the signal positivity condition in (3.2) (with J the period of the reference sinusoid) accurately reflects the local stability behavior of (3.1). Thus, if the algorithm were to reach those points of smaller d_*, local stability would result. Indeed, when initialized on the dc matching line with a d_* above 2.6, the motion of the parameter estimates was essentially along the dc matching line to parametrizations with d_* lower than 2.6 until halting near the $\text{avg}\{\phi_* e_*\} = 0$ setting described in Figure 6.7. This is in contrast to the case where the reference sinusoid frequency ω_r is 10. Here there is no stabilizing setting that satisfies the signal positivity condition even if $e_* = 0$. Thus, local stability should not be expected and was not observed.

Variations over time in θ_* due to changes in the plant description will also effect the error system as a disturbance that should remain small. Essentially, as shown in Anderson and Johnstone (1983), this can be interpreted as an upper bound on the frequency of these variations to keep e_* small. More subtle is the fact that variations in θ_* also represent variations in $H(z)$ and ϕ_*. Therefore, any variations in θ_* should be slow enough not to interfere with the total stability arguments. It can be shown that a nonzero rate of variation can be accommodated.

6.3.7 Frequency Domain Design Guidelines

An overview of the preceding primarily qualitative interpretation of the individual assumptions supporting satisfaction of our local stability theorems leads to the construction of Figure 6.11. Figure 6.11 accumulates a number of earlier insights into a suggested frequency spectrum of the bandwidths of various adaptive system components. This guideline is recommended by the theory of this book for robustness enhancement of adaptive controller applications. Satisfaction of this guideline requires a combination of selection and phrasing of an appropriate control task with prudent selection of designer-selected variables within the adaptive algorithm.

In the remainder of this section, we will briefly draw together statements that support the ordering of Figure 6.11. What will be noticed by experienced adaptive control practitioners is that many of

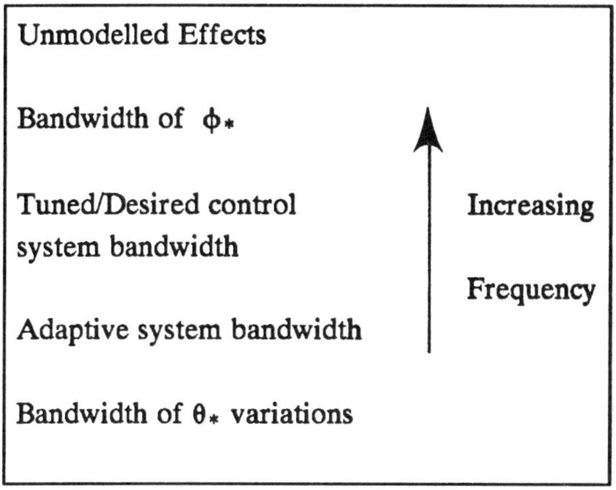

Fig. 6.11 Recommended frequency spectrum for robust adaptive control.

the aspects of this frequency response based guideline already reside in the "folklore" of adaptive control applications. One of the major contributions of the theory developed in this book is that it provides a detailed justification for and interpretation of this advisable design guideline.

The small step-size assumption is central to the averaging (or small gain theory) analysis dominating this book. Essentially, ϵ, or effectively the adaptive system convergence rate, should be chosen such that the variations in ϕ_* are much more significant to $H(z)$ than those in θ. This imposes a smaller lowpass "bandwidth" on (3.1), which is directly related to the adaptive system convergence rate, relative to the bandwidth of $H(z)$. In fact, this separation should be quite significant.

The tuned system bandwidth will necessarily be close to that of the desired transfer function in order to keep e_* small for regressor spectrum components above the tuned system bandwidth. As argued in Section 6.3.5, regressor spectrum components above the bandwidth of the tuned control system are needed to help boost the minimum eigenvalue of the signal positivity test. But regressor spectrum components above the bandwidth of $H(z)$ should be avoided to help the tradeoff in the signal positivity condition. This places the bandwidth of ϕ_* below that of $H(z)$ and above that of the tuned control system. Since the onset of unmodelled plant modes generates the break frequency of $H(z)$, this establishes the ranking of the top three terms in Figure 6.11.

To avoid insufficient averaging across the full bandwidth of the tuned control system behavior by the adaptive system in its determination of the average corrective action it will take, the bandwidth of the adaptive system should be well below that of the tuned system. The need for the separation between the bandwidth of θ_* variations and the adaptive system bandwidth was indicated in Section 6.3.6 as a consequence of maintaining a small e_*. This supports the lower portion of the hierarchy in Figure 6.11.

To help solidify the impact of this important guideline, summarized in Figure 6.11, the reader is encouraged to review the cxample of Section 6.2 to observe the adherence to this tenet.

6.4 ALGORITHM EXTENSION: REGRESSOR FILTERING

Following the publication in the open literature of examples of adaptive controllers exhibiting robustness difficulties came the proposal of a variety of algorithm modifications to cure these ills. The purpose of this section is to demonstrate the applicability of the analysis methods of this book to the substantiation of a demonstrably practical modification: regressor filtering. The resulting theoretical interpretation provides guidelines for prudent exploitation of an adaptive algorithm modification that should be considered as an adjunct to any adaptive control application. This theoretical substantiation of robustness improvement of a procedure based on sound engineering insight helps narrow the theory-practice gap in adaptive control engineering. Thus, this final section of the book is representative of the pragmatic advances we hope will emerge from the theory developed in this book and from that of subsequent efforts exploiting these same stability theory concepts.

Regressor filtering is based on the basic engineering concept of signal preconditioning. Assume that we have some practical information suggesting that adaptive system reaction to correction information due to a particular segment of the frequency spectrum of the regressor is unimportant or inadvisable. For example, the lowpass characteristic of control systems suggests that parameter estimate corrections due to high frequency regressor components should be suppressed. Thus, lowpass regressor filtering has been suggested as a reasonable adaptive controller modification, e.g. for the indirect adaptive controller of Chapter 5 in Johnson et al. (1984). This filtering takes place only on the regressor where it appears within the adaptive algorithm and not on its appearances within the control loop. Thus, the controller structure remains unaltered.

Recall that the algorithm form used, e.g., in Section 6.2 is

$$\hat{\theta}(k) = \hat{\theta}(k-1) + \frac{\epsilon}{1+\epsilon\phi^T(k-1)\phi(k-1)}\phi(k-1)\eta(k) \tag{4.1}$$

where

$$\eta(k) = d(k) - \phi^T(k-1)\hat{\theta}(k-1) \tag{4.2}$$

Note from (4.1) that

$$d(k) - \phi^T(k-1)\hat{\theta}(k) = \frac{\eta(k)}{1+\epsilon\phi^T(k-1)\phi(k-1)} \tag{4.3}$$

Using the tuned system configuration (as in (2.46))

$$d(k) - \phi^T(k-1)\hat{\theta}(k) = e(k) = H(z)[\phi^T(k-1)\theta(k)] + e_*(k) \tag{4.4}$$

where

$$\theta(k) = \theta_* - \hat{\theta}(k) \tag{4.5}$$

Thus, using (4.3) and (4.4) in (4.1), rewritten for θ via (4.5), yields the discrete-time version of the equation of interest in this book, as cited in (1.2) of Chapter 2 and (3.1).

Filtering of the regressor in *all* of its manifestations in the correction term of (4.1), including its necessarily supposed appearance in the desired signal $d(k)$, as indicated in (4.4), converts (4.1) to

$$\hat{\theta}(k) = \hat{\theta}(k-1) + \frac{\epsilon}{1+\epsilon\phi_F^T(k-1)\phi_F(k-1)}$$
$$\phi_F(k-1)[d_F(k)-\phi_F^T(k-1)\hat{\theta}(k-1)] \tag{4.6}$$

Given (4.4) and (4.5), (4.6) becomes

$$\theta(k) = \theta(k-1) - \epsilon\phi_F(k-1)\{H(z)[\phi_F^T(k-1)\theta(k)] + e_{F*}(k)\} \tag{4.7}$$

where

$$\phi_F(k) = F(z)[\phi(k)] \tag{4.8}$$

$$d_F(k) = F(z)[d(k)] \tag{4.9}$$

$$e_{F*}(k) = F(z)[e_*(k)]+(H(z)-1)\{F(z)[\phi^T(k-1)\hat{\theta}(k)]-\phi_F^T(k-1)\hat{\theta}(k)\} \tag{4.10}$$

As long as $F(z)$ has a sufficiently high bandwidth lowpass character relative to variations in $\hat{\theta}$, which are slowed by small ϵ, $e_{F*} \approx F(z)[e_*]$. Note that $H(z) \approx 1$ over the frequencies of $\phi_F^T\hat{\theta}$ and $F(z)[\phi^T\hat{\theta}]$, i.e., essentially ϕ_F, also enhances the accuracy of $e_{F*} \approx F(z)[e_*]$.

In certain cases the prediction error is *not* in the convenient $d(k)-\phi^T(k-1)\hat{\theta}(k-1)$ form of (4.1). For example, in the direct algorithm example of Section 6.2 the prediction error v in (2.19) [or in its $v(kT) = 0.2r((k-1)T)-y(kT)+0.9y((k-1)T)$ form] does not explicitly offer a separate addend of the form $\phi^T\hat{\theta}$. Compare this available form of the prediction error in the direct case to that of η

in (2.26) in the indirect algorithm example of Section 2. Here η is precisely in the form of (4.2). Though $d_F(k)-\phi_F^T(k-1)\hat{\theta}(k-1)$, as used in (4.6), could be fabricated, as $F(z)[v(k)+\phi^T(k-1)\hat{\theta}(k-1)]-\phi_F^T(k-1)\hat{\theta}(k-1)$, exploitation of the small step-size ϵ assumption is simpler. Then, under the assumption that $F(z)$ is a satisfactorily high bandwidth lowpass filter,

$$\eta_F(k) = F(z)[\eta(k)] = F(z)[d(k)-\phi^T(k-1)\hat{\theta}(k-1)] \approx d_F(k)-\phi_F^T(k-1)\hat{\theta}(k-1)$$

The ability to absorb the errors accumulating in the resulting redefinition of e_{F*} due to this approximation within the perturbation accommodated by the total stability theorem supports implementation of (4.6) as

$$\hat{\theta}(k) = \hat{\theta}(k-1) + \frac{\epsilon}{1+\epsilon\phi_F{}^T(k-1)\phi_F(k-1)}\,\phi_F(k-1)\eta_F(k) \quad (4.11)$$

where η_F is the filtered a priori prediction error, whatever its realization. Referral to the algorithms of Section 6.2 indicates that implementation of (4.11) simply requires prefiltering by $F(z)$ of all external signals entering the adaptive algorithm "box" and then computation of (4.1) as if the arriving quantities were unfiltered. This interpretation clearly exposes the signal conditioning basis of regressor filtering. As an interesting digression, consider filtering $\phi(k-1)$ by $F(z)$ and $\eta(k)$ by $G(z)$ in forming an alternative to (4.11). This is similar to the suggestion of (5.36) in Chapter 4. Essentially, this converts $H(z)$ and $e_{F*}(\approx F(z)[e_*])$ of (4.7) to $H(z)G(z)/F(z)$ and $e_{G*}(\approx G(z)[e_*])$, respectively. These redefinitions can be considered more flexible alternatives for $H(z)$ and e_{F*} in the following discussion.

From (4.7) it is apparent that the regressor filtering has three effects on the local stability indicators. In the first place, the positive definiteness requirement on a moving average of $\phi_* H(z)[\phi_*^T]$, as in (3.2), has been converted to one on $\phi_{F*}H(z)[\phi_{F*}^T]$. Secondly, the satisfactory degree of this positive definiteness is now relative to e_{F*}

rather than to e_*. Thirdly, the average convergence condition of $\text{avg}\{\phi_{F*}e_{F*}\} = 0$ replaces that of $\text{avg}\{\phi_* e_*\} = 0$. These three changes will now be examined to illustrate the potential benefit of appropriate regressor filtering in terms of robustness enhancement.

Clearly, the spectral distribution of ϕ_{F*} is different form that of ϕ_*. With $H(z) \approx 1$ at low frequencies, lowpass filtering ϕ_* enhances the strength of its low frequency spectrum relative to its high frequency content. This relative decrease in the "negative-weighted" energy of $\text{avg}\{\phi_* H(z)[\phi_*^T]\}$ may convert an $\text{avg}\{\phi_* H(z)[\phi_*^T]\}$ with a small negative eigenvalue to an $\text{avg}\{\phi_{F*} H(z)[\phi_{F*}{}^T]\}$ with all positive eigenvalues as desired. Thus, regressor filtering can be used to suppress the effect of too high a frequency tuned regressor component due to an unavoidable too-high frequency reference signal component (or disturbance).

Even if the average $\phi_* H(z)[\phi_*^T]$ is positive definite, its smallest eigenvalue may only be on the order of e_* and therefore the average $e_* \frac{\partial \phi_*}{\partial \theta_*}\big|_{\theta_*} + \phi_* H(z)[\phi_*{}^T]$ may not be positive definite. Recall the discussion in Section 6.3.6. Note that with regressor filtering this becomes a positive definiteness requirement on the average $e_{F*} \frac{\partial \phi_{F*}}{\partial \theta_*}\big|_{\theta_*} + \phi_{F*} H(z)[\phi_{F*}{}^T]$. If e_* is essentially a high frequency phenomenon, then e_{F*} $(\approx F(z)[e_*])$ will be attenuated relative to e_* by a lowpass $F(z)$ with unity dc gain. In terms of the counterbalancing force morphology used in interpreting the examples of Section 6.2, what may have been an apparently unhalted disturbance-direction drift before regressor filtering could now be halted around a stabilizing parametrization. Note that any spectral content of ϕ_* above the breakpoint of $F(z)$ will also be attenuated. So to keep the minimum eigenvalue from falling with the tuned prediction error, the dominating content of ϕ_* should not extend too far above the cutoff frequency of $F(z)$.

Exploitation of the potential benefits of these first two influences establishes a simple frequency domain guideline for lowpass regressor filter cutoff frequency selection. The regressor filtering cutoff frequency should be chosen in the vicinity of the frequency at which $H(z)$ acquires more than $\pm$ 90° phase shift. A cutoff well below this frequency can be supported given added information concerning the spectral distribution of the tuned regressor. For robustness improvement this cutoff should be well below the frequencies at which e_* has its principal content due, e.g., to any output measurement noise.

The redefinition of the average convergence condition to $\text{avg}\{\phi_{F*}e_{F*}\} = 0$ has the effect of enhancing the importance of reducing the prediction error within the passband of $F(z)$. With $F(z)$ a lowpass filter, this improves the low frequency fit of the adapted system to its objective. To support this claim heuristically recall the frequency response fitting paradigm. What regressor filtering does is suppress the observation of high frequency energy in the regressor (and prediction error). Thus, in fitting a frequency response curve from a parametrized low-order model through the observed magnitude gain and phase shift pairs at the component frequencies, the fit of points within the passband of $F(z)$ is given more weight than the fit of those in the stopband. This frequency weighting interpretation can be formalized, as in Anderson et al. (1985) for the addition of regressor filtering to the indirect adaptive controller of Chapter 5. There, as discussed in a continuous-time setting in Section 4.5, the fact used is that when ϵ is small, the tuned parametrization setting $\text{avg}\{\phi_* e_*\} = 0$ is approximately optimal in the sense that $\text{avg}\{(e_{F*})^2\}$ is effectively minimized. Thus, with regressor filtering it can be shown that effectively $\text{avg}\{(e_{F*})^2\}$ is minimized, which, given (4.10), indicates the frequency weighted optimization induced by regressor filtering.

Reconsider the prediction error generation diagrammed for the direct algorithm of Section 6.2 in Figure 6.5a. Here the prediction error is the model-following error filtered by the denominator polynomial $M_D(z)$ of the reference model transfer function $M(z) = M_N(z)/M_D(z)$. Thus, without regressor filtering, the prediction error being minimized is a highpass version of the model-following error. Note the resulting enhanced sensitivity to the very high frequency noise content in the model-following error that we presume is to be ignored. With regressor filtering by $F(z)$ as in (4.11), the prediction error being minimized is the model-following error filtered by $F(z)M_D(z)$. Thus, lowpass regressor filtering will tend to flatten out (and possibly roll off) the high frequency gain curve of $F(z)M_D(z)$ relative to that of $M_D(z)$ alone. In the frequency guidelines in Figure 6.11 the desired system (i.e., the reference model) bandwidth is less than that of the unmodelled effects. It is precisely in this interim range that the cutoff frequency of regressor filtering should vary. This achieves the previous suggestion of having the lowpass $F(z)$ most strongly attenuate the high frequency "negative-weighted" frequency range of $H(z)$. If the lowpass cutoff of the regressor filtering is above that of the reference model, then $|F(z)M_D(z)|$ will (likely) have a hump in the crossover frequency range of the achieved control system. Note that this will emphasize the minimization of the squared model-following error in those frequencies where good modelling for robust dynamic control design purposes is most important (Wittenmark and Astrom, 1984). Since the prediction errors for the indirect and direct schemes of Section 6.2 are a constant multiple of each other, these same remarks apply with the addition of regressor filtering to the indirect algorithm of Chapter 5. See Johnson et al. (1984) and Anderson et al. (1985) for an elaboration of similar remarks concerning the indirect scheme of Chapter 5.

As a particular example of the possible benefits of regressor filtering consider the direct algorithm example of Section 6.2. In that example, to counterbalance the e_* due to the disturbance sinusoid $0.04 \sin(20kT)$ (where $T = 0.05$), we added a reference sinusoid of greater magnitude, actually $0.1 \sin(\omega_r kT)$. This added sinusoid may be considered too large in its degrading effect on the fixed set-point tracking objective. We might even wish to lower the magnitude of the reference sinusoid to the order of the disturbance signal. Consider dropping the amplitude to 0.02, so the reference signal in (2.43) becomes

$$r(kT) = 2 + 0.02\sin(\omega_r kT) \qquad (4.12)$$

When ω_r is above the achieved bandwidth, it is hoped approximately 2, then the effect of the reference sinusoid on the desired 4 unit output is less than that of the "1% noise."

Applying the reference signal of (4.12) with $\omega_r = 6$ to the direct adaptive control via (2.15)-(2.20) of the plant in (2.1) with $g = 5$ results in a drift "down and to the left" along the dc matching line that appears unhalted in the stabilizing range of d. Note that this drift direction matches that when r is only a 2 unit step. This indicates that $\text{avg}\{e_* \frac{\partial \phi_*}{\partial \theta_*}\}$ is "more negative" than $\text{avg}\{\phi_* H(z)[\phi_*^T]\}$ is positive over the stabilizing range of d on the dc matching line. Thus, the minimum eigenvalue of $\text{avg}\{\phi_* H(z)[\phi_*^T]\}$, though positive, is insufficiently large. Recall Figure 6.10 in the preceding section, which showed that this same phenomenon occurred over only a portion of the stabilizing range of d along the dc matching line with the larger reference sinusoid signal (and $\omega_r = 8$).

Adding regressor filtering with

$$F(z) = \frac{(1 - \alpha)z}{z - \alpha} \qquad (4.13)$$

where $\alpha = 0.8$, halts the drift around $(c,d) = (1.7, 0.65)$ for an implied (a,b) of about (0.976, 0.118). Note that the regressor filter $F(z)$, which has its cutoff frequency around 4.5 rad/sec, is sandwiched between the desired bandwidth of 2 and the frequencies of the unmodelled effects (at approximately 10 for the unmodeled mode and 20 for the noise). This is compatible with the frequency response design guidelines suggested in Figure 6.11. Regressor filtering was added to the direct algorithm of (2.16)-(2.20) by first passing $r,y,$ and v through $F(z)$ before using the filtered versions in (2.16)-(2.20). Thus, (2.16)-(2.19) are replaced by

$$c(kT) = c((k-1)T) + \rho((k-1)T)r_F((k-1)T)v_F(kT) \tag{4.14}$$

$$d(kT) = d((k-1)T) - \rho((k-1)T)y_F((k-1)T)v_F(kT) \tag{4.15}$$

$$\rho((k-1)T) = \frac{0.01}{1+0.01(r_F^2((k-1)T)+y_F^2((k-1)T))} \tag{4.16}$$

$$v_F(kT) = x_F(kT) - y_F(kT) - 0.9[x_F((k-1)T)-y_F((k-1)T)] \tag{4.17}$$

where

$$r_F(kT) = (1-\alpha)r(kT) + \alpha r_F((k-1)T) \tag{4.18}$$

$$y_F(kT) = (1-\alpha)y(kT) + \alpha y_F((k-1)T) \tag{4.19}$$

$$x_F(kT) = (1-\alpha)x(kT) + \alpha x_F((k-1)T) \tag{4.20}$$

Unfiltered versions of y and r are still used to form the control signal via (2.15).

The (average) convergent (c,d) with the reference of (4.12) and regressor filtering through F(z) of (4.13) with $\alpha = 0.8$ is only modestly different from the convergent location of (1.84, 0.72) observed given the reference signal of (2.43) with the same reference sinusoid frequency $\omega_r = 6$ and 5 times the reference sinusoid amplitude but without regressor filtering. This suggests that the effect of the output measurement noise on the step tracking objective is about the same with or without regressor filtering in this example. However, as is expected, the maximum step tracking inaccuracy is significantly reduced due to the smaller amplitude reference sinusoid allowed with regressor filtering. In this same example, this lowpass regressor filtering will also beneficially suppress the higher frequency harmonics of anticipated (but infrequent) step changes in the set-point tracking level. Thus, regressor filtering is undeniably of practical benefit.

6.5 NOTES AND REFERENCES

A number of examples of adaptive controller destabilization due to unmodelled modes and output measurement disturbances in the absence of persistent excitation were provided in Egardt (1979). The misbehavior examples that received the most discussion are presented in Rohrs et al. (1985). A succinct, revealing commentary on the "Rohrs' " examples appears in Astrom (1983) and Astrom (1984). Compelling examples demonstrating the importance of the frequency content of regressor signals are provided in Ioannou and Kokotovic (1983).

The pragmatic appeal of signal preconditioning in adaptive control is routinely acknowledged, as indicated by remarks in Wittenmark and Astrom (1984). One of the first formalizations of this concept was in conjunction with the indirect adaptive control scheme of Chapter 5 and Section 6.2, as in Johnson et al. (1984) and Anderson et al. (1985).

REFERENCES

Anderson, B.D.O., and S. Vongpanitlerd, (1973), *Network Analysis and Synthesis: A State Space Approach,* Prentice-Hall, New Jersey.

Anderson, B.D.O., (1977), "Exponential stability of linear equations arising in adaptive identification," *IEEE Trans. on Aut. Contr.,* **AC-22**:83-88.

Anderson, B.D.O., and C.R. Johnson, Jr., (1982a), "Exponential convergence of adaptive identification and control algorithms," *Automatica,* **18**:1-13.

Anderson, B.D.O., and C.R. Johnson, Jr., (1982b), "On reduced-order adaptive output identification and adaptive IIR filtering," *IEEE Trans. on Aut. Contr.,* **AC-27**:927-933.

Anderson, B.D.O., and R.M. Johnstone, (1983), "Adaptive systems and time-varying plants," *Int. J. Contr.,* **37**(2):367-377.

Anderson, B.D.O., R.R. Bitmead, C.R. Johnson, Jr., and R.L. Kosut, (1984), "Stability theorems for the relaxation of strictly positive real condition in hyperstable adaptive schemes," *23rd IEEE Conf. on Dec. and Contr.,* Las Vegas, NV, 1286-1291.

Anderson, B.D.O., (1985), "Adaptive systems, lack of persistency of excitation and bursting phenomena," *Automatica,* **21**(3):247-258.

Anderson, B.D.O., R.R. Bitmead, C.R. Johnson, Jr., and R.L. Kosut, (1985), "Positive real conditions for adaptive control are not really necessary," *Proc. 7th IFAC Symp. on Ident. and Sys. Par Est.,* York, England 1003-1008.

Arnold. L., H. Crauerl, and V. Wihstutz, (1983), "Stabilization of linear systems by noise," *SIAM J. Contr. and Opt.,* **21**:451-461.

Astrom, K.J., (1983), "Analysis of Rohrs counterexamples to adaptive control," *Proc. 22nd IEEE Conf. on Dec. and Contr.,* San Antonio, TX, 982.

Astrom, K.J., (1984), "Interactions between excitation and unmodeled dynamics in adaptive control," *Proc. 23rd IEEE Conf. on Dec. and Contr.*, Las Vegas, NV, 1276-1281.

Bellman, R., (1953), *Stability Theory of Differential Equations*, McGraw-Hill, New York.

Benveniste, A., (1984), "Design of monostep and multistep algorithms for the tracking of time vary systems," *Proc. 23rd IEEE Conf. on Dec. and Contr.*, Las Vegas, NV, 353-358.

Bitmead, R.R., and B.D.O. Anderson, (1980a), "Lyapunov techniques for the exponential stability of linear difference equations with random coefficients," *IEEE Trans. on Aut. Contr.*, **AC-25**:782-787.

Bitmead, R.R., and B.D.O. Anderson, (1980b), "Performance of adaptive estimation algorithms in dependent random environments," *IEEE Trans. on Aut. Contr.*, **AC-25**:788-794.

Bitmead, R.R., (1984), "Persistence of excitation conditions and the convergence of adaptive schemes," *IEEE Trans. on Info. Thy.*, **IT-30**:183-191.

Bitmead, R.R., and R.K. Boel, (1985), "On stochastic convergence of infinite matrix products and its application to adaptive estimation," *Proc. 7th IFAC Symp. on Parameter Est. and Sys. Identification*, York, England, 1223-1228.

Bodson, M., S. Sastry, B.D.O. Anderson, I. Mareels, and R.R. Bitmead, (1986), "Nonlinear averaging theorems, and the determination of parameter convergence rates in adaptive control," *Systems and Contr. Letters*, to appear.

Boyd, S., and S.S. Sastry, (1983), "On parameter convergence in adaptive control," *Systems and Contr. Letters*, **3**:311-319.

Butchart, R.L., and B. Shackloth, (1965), "Synthesis of model reference adaptive control systems by Lyapunov's second method," *Proc. IFAC Symp. on Adaptive Contr.*, Teddington, England.

Coppel, W.A., (1965), *Stability and Asymptotic Behavior of Differential Equations,* D.C. Heath and Company, Boston, MA.

Desoer, C.A., and M. Vidyasagar, (1975), *Feedback Systems: Input-Output Properties,* Academic Press, New York.

Egardt, B., (1979), *Stability of Adaptive Systems,* Springer-Verlag, Berlin.

Fu, L., M. Bodson, and S. Sastry, (1985), "New stability theorems for averaging and their application to the convergence analysis of adaptive identification and control schemes," *Proc. 24th IEEE Conf. Dec. and Contr.,* Ft. Lauderdale, FL, 473-477.

Goodwin, G.C., and P.J. Ramadge, (1979), "Design of restricted complexity adaptive regulators," *IEEE Trans. on Aut. Contr.,* **AC-24**:584.

Goodwin, G.C., P.J. Ramadge, and P. E. Caines, (1980), "Discrete-time multivariable adaptive control," *IEEE Trans. on Aut. Contr.,* **AC-25**:449-456.

Goodwin, G.C., and K.S. Sin, (1984), *Adaptive Filtering, Prediction and Control,* Prentice Hall, New Jersey.

Green, M., and J.B. Moore, (1985), "Persistence of excitation in linear systems," *Proc. American Contr. Conf.,* Boston, 412-417.

Hahn, W., (1967), *Stability of Motion,* Springer-Verlag, Berlin.

Hale, J.K., (1980), *Ordinary Differential Equations,* Kreiger, Molaban, FL, originally published (1969), Wiley (Interscience), New York.

Ioannou, P.A., and P.V. Kokotovic, (1984), "Instability analysis and improvement of robustness of adaptive control," *Automatica,* **20**(5):583-594.

Ioannou, P.A., and P.V. Kokotovic, (1983), *Adaptive Systems with Reduced Models,* Springer-Verlag, New York.

Johnson, C.R. Jr., B.D.O. Anderson, and R.R. Bitmead, (1984), "A robust locally optimal model reference adaptive controller," *Proc. 23rd IEEE Conf. on Dec. and Contr.*, Las Vegas, NV, 993-998.

Kokotovic, P.V., B.D. Riedle, and L. Praly, (1985), "On a stability criterion for slow adaptation," *Systems and Contr. Letters*, **6**:7-14.

Kokotovic, P.V., H.K. Khalil, and J. O'Reilly, (1986), *Singular Perturbation Methods in Control: Analysis and Design*, Academic Press, New York.

Kosut, R.L., and B. Friedlander, (1982), "Performance robustness properties of model reference adaptive control," *Proc. 21st IEEE Conf. Dec. and Contr.*, Orlando, FL.

Kosut, R.L., and C.R. Johnson, Jr., (1984), "An input-output view of robustness in adaptive control," *Automatica*, **20**(5):569-581.

Kosut, R.L., B.D.O. Anderson, and I.M.Y. Mareels, (1985), "Stability theory for adaptive systems: methods of averaging and persistency of excitation," *Proc. 24th IEEE Conf. Dec. and Contr.*, Ft. Lauderdale, FL, 478-483;. *IEEE Trans. on Aut. Contr.*, to appear.

Kosut, R.L., and B. Friedlander, (1985), "Robust adaptive control: conditions for global stability," *IEEE Trans. on Aut. Contr.*, **AC-30**(7):610-624.

Kosut, R.L., and B.D.O. Anderson, (1986), "Local stability analysis for a class of adaptive systems," *IEEE Trans. on Aut. Contr.*, **AC-31**(1):86-89.

Landau, Y.D., (1976), "Unbiased recursive identification using model reference adaptive techniques," *IEEE Trans. Aut. Contr.*, **AC-21**: 194-202.

Landau, I.D., (1979), *Adaptive Control: The Model Reference Approach*, Marcel Dekker, Inc., New York.

Ljung, L., (1977), "On Positive real transfer functions and the convergence of some recursions," *IEEE Trans. on Aut. Contr.*, **AC-22**:539-551.

Ljung, L., and T. Soderstrom, (1983), *Theory and Practice of Recursive Identification,* MIT Press.

Ljung, L., (1985), "Asymptotic variance expressions for identified black-box transfer function models," *IEEE Trans. on Aut. Contr.*, **AC-30**:834-844.

Luenberger, D.G., (1969), *Optimization by Vector Space Methods,* John Wiley and Sons, Inc., New York.

Mareels, I.M.Y., (1984), "Sufficiency of excitation," *Systems and Contr. Letters,* 159-163.

Mareels, I.M.Y., and R.R. Bitmean (1985), "Operator decomposition of two-time scale systems,{{ Technical Report, Dept. of Systems Engineering, Australian National University.

Meerkov, S.M., (1980), "Principle of vibrational control: theory and applications," *IEEE Trans. on Aut. Contr.*, **AC-25**:755-762.

Miller, R.K., and A.N. Michel, (1982), *Ordinary Differential Equations,* Academic Press, New York.

Monopoli, R.V., (1974), "Model reference adaptive control with an augmented error signal," *IEEE Trans. on Aut. Contr.*, **AC-19**:474-484.

Morgan, A.P., and K.S. Narendra, (1977), "On the stability of nonautonomous differential equations $\dot{x} = [A + B(t)]x$ with skew symmetric matrix $B(t)$," *SIAM J. Contr. and Opt.*, **15**:163.

Narendra, K.S., and J.H. Taylor, (1973), *Frequency Domain Criteria for Absolute Stability,* Academic Press, New York.

Narendra, K.S., Y.H. Lin, and L.S. Valavani, (1980), "Stable adaptive control design, part II: proof of stability," *IEEE Trans. on Aut. Contr.*, **AC-25**(3):440-449.

Ortega, R., L. Praly, and I.D. Landau, (1985), "Robustness of discrete-time direct adaptive controllers," *IEEE Trans. on Aut. Contr.*, **AC-30**:1179-1187.

Ortega, R., and R. Kelly, (1985), "On the destabilizing effect of cross-coupling in decentralized adaptive control," *Systems and Contr. Letters*, **6**:99-103.

Parks, P.C., (1966), "Lyapunov redesign of model reference adaptive control systems," *IEEE Trans. on Aut. Contr.*, **AC-11**:362.

Praly, L., (1985a), "A geometric approach for the local analysis of a one step ahead adaptive controller," *Proc. 4th Yale Workshop on Applications of Adaptive Systems Theory*, Yale University, CT.

Praly, L., (1985b), "Local stability-instability of slowly adapted systems in presence of periodic inputs," Technical Report, CAI, Ecole des Mines.

Praly, L., and D. Rhode, (1985), "A local analysis of a one step ahead adaptive controller," *Proc. 24th IEEE Conf. on Dec. and Contr.*, Ft. Lauderdale, FL, 1862-1867.

Riedle, B.D., and P.V. Kokotovic, (1984), "A Stability-instability boundary for disturbance-free slow adaptation and unmodeled dynamics," *Proc. 23rd IEEE Conf. on Dec. and Contr.*, Las Vegas, NV, 998-1001.

Riedle, B.D., and P.V. Kokotovic, (1985a), "Stability analysis of an adaptive system with unmodelled dynamics," *Int. J. Contr.*, **41**:389-402.

Riedle, B.D., and P.V. Kokotovic, (1985b), "A Stability-instability boundary for disturbance-free slow adaptation and unmodeled dynamics," *IEEE Trans. on Aut. Contr.*, **AC-30**:1027-1030.

Riedle, B.D., and P.V. Kokotovic, (1985), "Integral manifold approach to slow adaptation," *24th IEEE Conf. on Dec. and Contr.*, Ft. Lauderdale, FL.

Riedle, B.D., L. Praly, and P.V. Kokotovic, (1986), "Examination of the SPR condition in output error parameter estimation," *Automatica,* to appear.

Rohrs, C.E., L.S. Valavani, M. Athans, and G. Stein, (1985), "Robustness of continuous-time adaptive control algorithms in the presence of unmodeled dynamics," *IEEE Trans. on Aut. Contr.* **AC-30**:881-889

Volosov, V.M., (1962), "Averaging in systems of ordinary differential equations," *Russian Math. Surveys,* **17.**

Wittenmark, B., and K.J. Astrom, (1984), "Practical issues in the implementation of self-tuning control," *Automatica,* **20**:595.

Yoshizawa, T., (1975), *Stability Theory and the Existence of Periodic Solutions and Almost Periodic Solutions,* Springer-Verlag, Berlin.

Yuan, J. S-C. W.M. Wonham, (1977), "Probing signals for model reference identification," *IEEE Trans. on Aut. Contr.,* **AC-22**:530.

Zames, G., (1966), "On the input-output stability of time-varying nonlinear feedback systems," *IEEE Trans. on Aut. Contr.,* Part I: **AC-11**(2):228-238, Part II: **AC-11**(3):465-476.

INDEX